# Local Forest Management

## The Impacts of Devolution Policies

*Edited by*
David Edmunds
and
Eva Wollenberg

London • Sterling, VA

First published in the UK and USA in 2003
by Earthscan Publications Ltd

Reprinted 2004

ISBN: 1-84407-023-9  paperback
       1-84407-022-0  hardback

Typesetting by JS Typesetting Ltd, Wellingborough, Northants
Printed and bound in the UK by Creative Print and Design (Wales), Ebbw Vale
Cover design by Danny Gillespie

For a full list of publications please contact:

Earthscan
8–12 Camden High Street, London, NW1 0JH, UK
Tel: +44 (0)20 7387 8558
Fax: +44 (0)20 7387 8998
Email: earthinfo@earthscan.co.uk
Web: **www.earthscan.co.uk**

22883 Quicksilver Drive, Sterling, VA 20166-2012, USA

A catalogue record for this book is available from the British Library

Library of Congress Cataloging-in-Publication Data

Local forest management : the impacts of devolution policies / edited by David
Edmunds and Eva Wollenberg.
        p. cm.
    Includes bibliographical references (p. ).
    ISBN 1-84407-023-9 (pbk.) – ISBN 1-84407-022-0 (hardback)
        1. Forest policy–China. 2. Forest policy–India. 3. Forest policy–Philippines.
    I. Edmunds, David. II. Wollenberg, Eva

    SD643.L63 2003
    333.75'095–dc21

                                                                        2003008345

Earthscan publishes in association with WWF-UK and the International Institute for
Environment and Development

This book is printed on elemental chlorine free paper

# Contents

# List of Tables and Boxes

## TABLES

## BOXES

# Foreword

This book is the output of a three-country study on creating space for local forest management, initiated and supported by the International Fund for Agricultural Development (IFAD) through the Center for International Forestry Research (CIFOR). Over 30 partners in five Asian countries contributed to the study. The results of this unique exercise have profound implications both for the sustainable management of forests, as well as for ensuring sustainable livelihoods for poor forest-dependent communities. Drs Phrang Roy and Shantanu Mathur of IFAD point out in their thought-provoking Preface that sustainable livelihood opportunities are essential for the poor if they are to take an interest in forest conservation. In other words, there has to be an economic stake in conservation. Conservation of natural resources and poverty eradication will then become mutually reinforcing.

The poor are poor mainly because they have no assets such as land, livestock, fish ponds or special skills. They survive on uncertain daily wages, carrying out unskilled jobs. Under such conditions, the poor can come out of the poverty trap only through asset-building and human resource development. The studies reported in this book were carried out in China, India and the Philippines. They reveal both the maladies responsible for the poverty of communities living in, or in the vicinity of, forests, as well as the potential remedies. The Poverty Eradication Compact should include decentralization of forest management so as to achieve a shift in authority to the poorest forest-dependent communities; re-distributive land reform; the involvement of women in all phases of sustainable forest management (for example, conservation, sustainable use and equitable benefit-sharing); and the introduction of multiple sources of income, including, where appropriate, eco-tourism and enterprises based on non-timber forest products. The book contains information on successful systems of sustainable and equitable management of forest resources. Work done in India and Nepal on under-utilized crops such as nutritious millets, legumes and tubers, with support from IFAD, has also shown that both the nutrition and livelihood security of the poor living near forests can be strengthened by reviving dying wisdom and dying crops. Furthermore, there are many sustainable livelihood options for resource-poor families living near mangrove forests in the coastal areas of India, China and the Philippines.

This book is a fine example of the most meaningful work that IFAD has stimulated and supported during the last 25 years for saving nature, as well as lives and livelihoods. IFAD has demonstrated how the *antyodaya* (beginning with the poorest person) approach to development advocated by Mahatma Gandhi is the most effective pathway for fostering high social synergy and a win–win pattern of sustainable development.

We owe a deep debt of gratitude to Drs Phrang Roy and Shantanu Mathur and to the contributors to this book for this labour of love in the cause of achieving a poverty-free world.

M S Swaminathan

# Preface

Poverty in rural Asia is more prominent among communities who live in, or in the vicinity of, forests. The forest poor include many indigenous peoples who have to endure factors external to their natural resource base that entrench and perpetuate their poverty. These externalities deprive them of opportunities to fully utilize their resources and, often inert, capacities, and impede their agency to improve their own livelihood security.

The forest poor inhabit areas that lack basic infrastructural amenities, such as piped water supply, sanitation and electricity. Their access to financial services, inputs and technology is severely limited. The constraints also include a general lack of an enabling institutional architecture that can better position the rural poor to derive the best terms from the sustainable utilization of their natural resource base. The institutional disadvantages include lack of information about markets, lack of business and negotiating experience and, particularly, lack of a strong collective organization – depriving the poor of effective participation and the power needed to interact on equal terms with the other, generally larger and stronger, players in the market, including intermediaries. Cultural and social distance, and discrimination are some of the other institutional factors that also, at least partly, exclude the poor from markets. Low levels of social and physical infrastructure increase their vulnerability to famine and disease, especially in the mountainous and remote areas of the region.

The International Fund for Agricultural Development's (IFAD's) mission is to achieve the greatest possible impact in enabling the rural poor to overcome their poverty. We have found that poverty is not just multidimensional. Poverty is a condition where individuals have been deprived of their capabilities to lead the kind of lives they value, to be free of fear and to be able to express themselves. This capability approach, pioneered by Amartya Sen, underlies our strategies to reduce poverty. Poverty reduction, then, is not only a matter of service delivery. It is about finding ways of enhancing the agency of the poor, of women and of men, to transform their production capabilities and their lives.

In addressing poverty in Asia, IFAD seeks to address the *structural causes* of the lack of agency of the poor. This entails, for the fund, the challenge of tackling the causes of restricted access to productive resources, and actually increasing control over the use of these resources. Changes in access to resources – such as property reform for forests in the uplands, access to finance and other resources, and land reform through land distribution or greater access rights to the landless – will increase the income and overall capability achievement of the poor. It will also contribute to increasing production and productivity in their respective local and national economies. Such changes entail redistribution strategies that enhance productivity – essential if poverty reduction is to be sustained in an age of globalization.

In forest areas, a central issue, then, is secure and reasonable property rights over the forests, which are the productive assets of the majority of the indigenous people and other marginalized groups of rural poor. In recent years, there has been an increasing trend towards devolution of control over natural resources from central government to local communities. The emphasis of such devolution has been the sustainability of resources to be used by all, rather than poverty reduction through securing livelihoods for the poor.

To better understand the devolution of forest management and its implications for access to productive resources for the forest poor, in 1998 IFAD provided grant support to the Center for International Forestry Research (CIFOR) for a three-year study on 'Creating Space for Local Forest Management' in Asia. CIFOR coordinated this excellent piece of research, conducted with over 30 partners in five countries in Asia, ably led by David Edmunds and Eva Wollenberg and a team that included Madhu Sarin, Neera Singh, Liu Dachang, Antonio Contreras, Dev Nathan, Govind Kelkar, Jeff Fox, Louise Buck and Jon Anderson, who contributed both to the analytical rigour, as well as to the careful nuancing of the community-based research agenda that becomes necessary when dealing with socio-cultural and economic issues relating to the diversity of the forest poor. The study aimed to understand the impact of devolution policies in India, China and the Philippines and to explore ways of empowering local communities under these policies. Together with their partners, CIFOR produced more than 50 publications, including this book, as one element of the dissemination of their findings.

The main finding of the project was that seemingly constructive and favourable devolution policies have a rather mixed pro-poor impact. Counterintuitively, such decentralization does not always produce a genuine shift in authority to the poorest forest users. Although there have been incremental gains for many local forest users, the policies have also enabled forest departments to control forests in new ways. For forest users who already enjoyed substantial control over and benefits from forests, devolution policies facilitated the extension of state control where it did not exist before. For forest users who previously lacked access to forest products or land and sometimes gained significant benefits from devolution policies, these gains were constrained by bureaucratic controls and elite power structures. The emphasis of devolution has been on transferring responsibilities for planting and protecting to forest users, while retaining authority over most aspects of management. By supervising or organizing planning processes, local organizations or tax and regulatory incentives, forest departments exerted control at far less cost to themselves, and could afford to reach into forests where they had no effective presence before. As a result, forest departments maintained or even increased their influence over forests with devolution policies.

The research reinforces the experience of many IFAD loan-financed projects that land reform – both tenancy reform and redistribution of ceiling surplus lands to the landless – is important to poverty reduction. In addition to production benefits, land reform helps to change the local political structure by giving more voice to the poor. Re-distributive land reform, whether through market-assisted land reform programmes or otherwise, should remain a substantive policy issue for poverty reduction.

In addition, the study reinforces our observation that institution-building must be just, genuinely participatory and inclusive to benefit the poor. If the poor are to benefit from cooperative arrangements for resource management, they must participate fully in setting them up. Such arrangements can be built on the basis of patron–client relationships; but these are unlikely to advance the living standards of the poor much beyond the level of survival. The alienation of the clients from the residual income accruing from their labour makes it difficult and costly, even impossible, to monitor their performance effectively. This alienation within patron–client relations would reduce productivity itself, while a more inclusive situation is likely to enhance productivity.

As traditional hierarchies are being eroded everywhere with the spread of education and democracy, a new basis of cooperation must be created so that the appropriators can themselves devise viable institutions by mutual consent. Even in this case, however, the poor might be excluded. It is essential that the rights of the poor to participate in the decision-making process are protected, promoted and upheld. Today, this is a fundamental requirement of most IFAD-funded projects.

Participation is, however, insufficient by itself. It needs to be built upon, improving the livelihood security of the poor. The poor who live at the edge of subsistence necessarily place a high value on their time: if conservation of natural resources comes into conflict with meeting the immediate need for survival, then they will disregard the concern for conservation.

Women, in particular, directly bear the cost of poor participation in community institutions designed for natural resources management. Village forest protection committees regulate forest access for its sustainable use. The cost of foregoing forest use is disproportionately borne by the women through a larger amount of time spent in fuelwood collection. Women's collection time increased from between one and two hours to between four and five hours for a headload of firewood, and journeys of 0.5 kilometres lengthened to 8 or 9 kilometres soon after the start of protection in the Indian states of Gujarat and West Bengal.

Yet, the various Joint Forest Management (JFM) and Community Forest Management (CFM) initiatives show that where women and the poor do organize, they are able to make some impact on the agendas of the community institutions in ways that help them to meet their livelihood priorities. In Meghalaya, for instance, the strongly male-dominated village assembly, after continuous pressure from the IFAD project, has begun to include women in its deliberations. Women's issues of access to fuelwood and fodder have also been included within the forest management programmes.

The inclusion of women and other poor does not come about as a matter of course. Changing age-old traditions is a matter of struggle to change gender- or caste-based norms of behaviour – a struggle involving mainly the local participants, but also facilitated by external agencies, including project agencies, non-governmental organizations (NGOs) and government legislation and orders. The various laws and rules in different South Asian countries, including, most recently, Pakistan, that mandate a certain proportion of women members in village committees create a strong climate in favour of women's public and community roles. While such representation of women often starts out as nominal, over time it also tends to become more real and effective, but, of course, not without struggle.

Given the mixed record of devolution in Asia and the lack of genuine control by forest users of forests and their livelihoods, the CIFOR study suggests several directions for future development efforts. Firstly, poverty reduction for forest communities needs to move beyond devolution pro- grammes coordinated by forest departments. These efforts should seek to complement devolution programmes of forest departments by focusing on expanding local communities' influence and creating significant livelihood opportunities that are independent of government forest policy. They are likely to be most effective where they can identify viable income opportunities for local communities and, at the same time, enhance communities' capacities to self-organize, acquire and apply legal literacy, as well as building coalitions with other government bodies, non-government actors and other communities to support their interests. These efforts, in turn, should enable communities to mobilize resources and make demands to engage in more meaningful econ- omic development.

Secondly, any activity in the remaining forested areas of Asia necessarily demands the accommodation of multiple interests. No forest is any longer a single stakeholder affair. More attention needs to be given to fostering multi- stakeholder processes that are more effective in representing disadvantaged communities' views in accountable ways and negotiating their interests. Support should be provided for processes that develop shared frameworks about the aims of forest management and the mode of its implementation to achieve a more just balance of public and local interests. Encouraging joint learning among stakeholders is likely to enable better collaboration and adaptation. More capacity-building is needed to enable communities to participate in such processes strategically and to use them as only one of several means for meeting their needs.

Thirdly, more attention needs to be given to building support for indigenous perspectives and the management systems that local forest communities have developed. There is a need to incorporate forest users' perspectives in the design and evaluation of devolution programmes. In the same way that many protected areas around the world now include cultivated or human-modified landscapes and acknowledge the value of these landscapes, so there must be advocacy to recognize the value of the different kinds of forests that local communities manage. These classifications of forest management must also be put 'on the map'. Such advocacy cannot be based on pointing to a few examples of community management, which has been the case to date. Instead, there needs to be more fundamental recognition of the different values of indigenous groups and the development of methodologies that incorporate these values. The development of new policies and organizations within government and among NGOs is necessary to promote these perspectives.

The year 2003 is the 25th anniversary year of IFAD, and we are pleased that our partnership with CIFOR has led to a valuable contribution to knowledge and insights in community-based natural resource utilization and management – in this case, forest resources. We hope very much that this book – which attempts to raise some key questions regarding people-centered approaches to forest management and provides strong practical guidance for refining our thinking on the subject – will also help to better inform rural poverty-reduction initiatives in this vital sector.

Phrang Roy
Assistant President
External Affairs Department
IFAD

Shantanu Mathur
Grants Coordinator
Project Management Department
IFAD

# List of Acronyms and Abbreviations

| | |
|---|---|
| ACF | Assistant Conservator of Forests |
| ADMP | Ancestral Domains Management Programme |
| APRS | Agricultural Production Responsibility System (China) |
| BILT | Ballarpur Industries Ltd |
| BJFPC | Budhikhamari Joint Forest Protection Committee |
| CADC | Certificate of Ancestral Domain Claims |
| CBFM | Community-based Forest Management |
| CBFMA | Community-based Forest Management Agreements |
| CEP | Coastal Environmental Programme |
| CF | Conservator of Forests |
| CFM | Community Forest Management |
| CFSA | Community Forest Stewardship Agreements |
| CIFOR | Center for International Forestry Research |
| CPEU | Centers for People Empowerment in the Uplands |
| CRMP | Community Resources Management Programme |
| CSC | Certificate of Stewardship Contracts |
| CTF | Communal Tree Farm Programme |
| DC | Deputy Commissioner |
| DCF | Deputy Conservator of Forests |
| DENR | Department of Environment and Natural Resources |
| DFID | Department for International Development (UK) |
| DFO | Divisional Forest Officer |
| DPF | Demarcated Protected Forest |
| DRDA | District Rural Development Agency |
| EDC | Eco-Development Committee |
| FAR | Family Approach to Reforestation Programme |
| FCA | Forest Conservation Act, 1980 |
| FLMP | Forest Land Management Programme |
| FOM | Forest Occupancy Management Programme |
| FPC | Forest Protection Committee |
| FSP | Forest Sector Programme |
| GAD | gender and development |
| GOI | Government of India |

| | |
|---|---|
| GOO | Government of Orissa |
| GOUP | Government of Uttar Pradesh |
| GS | *gram sabha* |
| IFA | Indian Forest Act, 1927 |
| IFAD | International Fund for Agricultural Development |
| IFMA | Integrated Forest Management |
| IPRA | Indigenous Peoples Rights Act (the Philippines) |
| IRMP | Integrated Rainforest Management Program |
| ISF | Integrated Social Forestry |
| ISFP | Integrated Social Forestry Program |
| JFM | Joint Forest Management |
| LGUs | local government units |
| LIUCP | Low-Income Upland Communities Project |
| MFP | minor forest produce |
| MMD | Mahila Mangal Dal (Women's Welfare Association) |
| MOEF | Ministry of Environment and Forests (GOI) |
| MoF | Ministry of Forestry (China) |
| MoU | memorandum of understanding |
| MP | Madhya Pradesh State |
| MTOs | mass tribal organizations |
| NFP | National Forestation Programme |
| NGO | non-governmental organization |
| NIPAS | National Integrated Protected Areas Systems |
| NTFP | non-timber forest product |
| PA | protected area |
| PAMB | Protected Area Management Board |
| PCCF | Principal Chief Conservator of Forests |
| PCSD | Palawan Council for Sustainable Development |
| PESA | Panchayats (Extension to Scheduled Areas) Act, 1996 |
| PRI | Panchayati Raj (local self-governance) institutions |
| RCDC | Regional Centre for Development Communication |
| RD | Revenue Department |
| RF | Reserve Forest |
| RMB | *reminbe* (Chinese currency) |
| RRDP | Rain-fed Resources Development Programme |
| RRMP | Regional Resource Management Programme |
| SC | scheduled caste |
| SHT | spearhead team |
| SPWD | Society for the Promotion of Wasteland Development |
| UP | Uttar Pradesh |
| UPF | undemarcated protected forest |

| | |
|---|---|
| VC | village committee (a generic term used for VPs, VFCs and proposed JFM committees under the new draft Village Forest JFM rules) |
| VF | Village Forest |
| VFC | Village Forest Committee (formed under the Gram Panchayat Act where there is no VP) |
| VFJM | Village Forest Joint Management |
| VFJMR | Uttar Pradesh Village Forests Joint Management Rules |
| VP | *van panchayat* |
| WBFP | World Bank Forestry Project |
| WINDS | Women in Development of Sarangani (Philippines) |
| WPA | Wildlife Protection Act, 1972 |

# Glossary of Local Terms

| | |
|---|---|
| *agarhias* | name of a higher caste group, Orissa |
| *begar* | system requiring free labour |
| *benaap* | unmeasured |
| *bhawan* | building |
| *chipko* | (literally) sticking to or hugging |
| *datu* | customary leaders in southern areas of the Philippines |
| *gram sabha* | term used for the elected representatives of the lowest local government institution in Uttar Pradesh; in most other Indian states it means the general body of electors of a *gram panchayat* |
| *gramya jungles* | village forests |
| *haq haquque* | legal rights |
| *harijans* | name of a lower caste group, Orissa |
| *jungle manch* | forest forum |
| *jungle sarpanches* | forest president |
| *khesra* | local term for a category of revenue forest in Orissa |
| *lakh* | an Indian unit for 100,000 |
| *lath panchayat* | literally 'stick *panchayat*' – traditional institution in Uttarakhand which also protects and manages community forests |
| *mahila mandal* | women's association |
| *mahila samities* | women's organizations |
| *naap* | measured |
| *nyaya* | justice |
| *paryavaran* | environment |
| *patwari* | lowest-level revenue official |
| *pradhan* | head of *gram panchayat* in Uttarakhand |
| *reminbe* | Chinese currency |
| *rupee* | Indian currency |
| *Sal assi* | 80th year (according to Indian calendar) |
| *samitis* | organizations/committees |
| *sarpanch* | head of *van panchayat* (Uttarakhand) or *gram panchayat* (most other states) |
| *soyam* | term used for revenue forest land in the princely state of Tehri Garhwal |
| *tambon* | local government district, Thailand |
| *tehsil* | an administrative unit within a district |
| *van dhan samiti* | NTFP marketing committee |
| *van panchayat* | literally 'forest *panchayat*' – democratic community forest management institution in Uttarakhand |
| *van samiti* | forest committee |
| *vana samrakhan samitis* | literally 'Forest Protection Committees' – name of JFM committees in Orissa |
| *yuan* | Chinese currency |
| *zila parishad* | elected district government institution |

# 1

# Introduction

*David Edmunds, Eva Wollenberg, Antonio P Contreras,*
*Liu Dachang, Govind Kelkar, Dev Nathan,*
*Madhu Sarin and Neera M Singh*

During the 1980s, a confluence of political pressures began to force or encourage central governments to devolve natural resource management to local individuals and institutions located within and outside of the government:[1]

- Overextended government bureaucracies began to look for ways to cut costs.
- Environmentalists painted images of sustainable resource management based on an intimate economic and cultural connection between local people and natural resources, as well as images of more effective resource protection by those living in close proximity to natural resources.
- The poor and their advocates hoped that local control would help them to protect local livelihoods and capture a greater share of the other benefits of natural resource management.
- Development specialists demonstrated the feasibility of working with local communities, and an ideological movement was developed that supported more small-scale, bottom-up and locally responsive measures based on local people's self-determination, in contrast to development strategies focused on large, imposed infrastructural investments.
- Political reformers argued that direct public involvement in resource management and greater public oversight of (more accessible) local officials were ends in themselves, and that such decentralization improved civic culture.

These forces have encouraged one of the most dramatic transformations in natural resource management in modern history. There are now devolution policies for natural resources in virtually every corner of the globe. Policy reforms have sought to transfer authority for managing water resources to local institutions in more than 25 countries (Vermillion, 1997). Rights over wildlife have been devolved in Namibia, Zambia, Zimbabwe and Botswana, and rights over rangelands have been devolved to local management associations in Lesotho (Shackleton et al, 2002). Central governments have also taken steps to transfer authority for managing fisheries, soils, protected areas and other resources to local institutions.

Perhaps the most far-reaching and well-documented devolution policies, however, have been in forest management (Fisher, 1999; D'Silva and Nagnath, 2002, Lynch and Talbot, 1995, Poffenberger, 1990). Colonial regimes in Asia and Africa, in particular, centralized the management of forests to control the valuable resources found there (Guha, 2001) and, sometimes, the people who might be hiding among them (Sioh, 1998). Independent governments also began centralizing the management of forests during the 20th and early 21st centuries. Centralized state control continued throughout most of the 21st century.

However, governments found it increasingly difficult to exercise their authority effectively.[2] The world's remaining forests were remote, their resources diverse, and their interactions with people and other resources increasingly complex. Popular protest at the shortcomings of centralized policies, such as the rubber tappers' *'empate'* movement in Acre and the *chipko andolan* forest protection movement in India, also forced government officials in some countries to re-evaluate their position on who should be responsible for forest management.

Though the policies take different forms, the shift in forest management authority from central government has occurred around the globe. There are now policies to devolve forest management to municipalities in Bolivia (Kaimowitz et al, 2003), district councils in Zimbabwe (Mandondo, 2000; Nemarundwe, 2001) and Tanzania (Massawe, 2001; Wiley, 1997), and other forms of local government in Indonesia (Wollenberg and Kartodihardjo, 2002; Dermawan and Resosudarmo, 2002) and the Philippines (Magno, 2001). In India (Saxena, 1997), Nepal (Kafle, 2001; Shrestha and Britt, 1997; Malla, 2001) and the US and Canada (Poffenberger, 1995), central governments have granted authority for forest management to local non-governmental and community-based organizations. In China, individual households have benefited from devolution policies, exercising management rights once reserved for government or party-dominated collectives (Liu, 2001).

The shift in authority for forest management is formalized in a variety of ways, including:

- corporate legal organizations, composed of rights holders – for example, rubber tappers' organizations (Brazil), *ejidos* (Mexico), trusts (Botswana), conservancies (Namibia) and communal property associations (Makuleke, South Africa);
- village committees facilitated by government departments – for example, Village Natural Resource Management Committees in Malawi and Forest Protection Committees in India;
- contractual agreements between the government and households or individuals (the Philippines, China);
- local government organizations, such as rural district councils in Zimbabwe and *panchayats* in India, and multi-stakeholder district structures aligned to line departments, such as *tambon* councils in Thailand and wildlife management authorities in Zambia.

Some state programmes have also allocated rights and responsibilities directly to households or individuals, such as in China and the Philippines, where individuals exercise varying degrees of authority over species selection, harvesting practices, sale and consumption, and the distribution of benefits.

Many of these arrangements parallel those for other natural resources. In fact, local organizations created under devolution policies are often responsible for managing multiple resources – for example, district councils in Zimbabwe are involved in wildlife management, and local governments are responsible for protected areas in the Philippines. Because the motivations and formal arrangements for forest devolution are similar to those for other natural resources, we suggest that the lessons learned from forestry may be broadly applicable to the devolution of other natural resources.

Among the most important lessons, we believe, are those concerning the impact of devolution policies on the lives of the rural poor. A broadly held expectation – indeed, a key rationale for devolution policies – has been that devolution would bring the large numbers of rural poor who live in and near forests better access to forests and more self-determination in decisions about local resources. The results have been mixed and difficult to discern. Some observers suggest that, although devolution policies fail to deliver on many of our hopes, they are a better alternative than centralized management (Chiong-Javier, 1996; Corbridge and Jewitt, 1997; Joshi, 1999; Saigal, 2000; Lynch and Talbot, 1995; Marco and Nuñez, 1996). In a sense, some participation is better than none. Most governments and donors, encouraged by such analysis, have continued their efforts to devolve forest management away from central forest departments.

However, experience with specific policies suggests a need to pause and reflect about the nuances of devolution's impacts. Indeed, forest cover has increased in many places (Saxena, 1997; Liu, 2001). Policies have encouraged legitimate community tenure rights to forests (Lynch, 1999) and have helped to promote participatory decision-making in forest management (Johnson, 1999; Guha, 2001). Yet, in many cases, devolution appears to be transferring little or no authority to local forest users and is having, at best, no significant positive impact on the livelihoods of the poor. Local institutions set up under devolution have often been accountable to forest departments and other government offices, rather than to local people (Ribot, 1998; Malla; 2001; Mandondo, 2000), with the possibilities for genuine co-management being quite limited (Baland and Platteau, 1996) despite efforts to empower local communities under these arrangements (Borrini-Feyerabend, 2000). Devolution has not proportionately benefited women, ethnic minorities or the very poor; any gains in income have been relatively small for most people, and often overshadowed by negative trade-offs in resource access and control (Upreti, 2001; Sarin, 1998). Pre-existing local institutions have been undermined by their lack of legal standing (Lindsay, 1998) and clear property rights (Grinspoon, 2001; Agrawal and Ostrom, 2001) relative to institutions that are newly created or sponsored by governments.

To better understand the nuances of these impacts of devolution on the poor, we decided that a study of devolution's impacts was needed from a new perspective. Although a number of studies have documented the mixed impacts of devolution on local people's livelihoods and political power in specific sites (Poffenberger, 1995; Sarin, 1998; Khare et al, 2000; Sundar, 2000a; Gauld, 2000; Malla, 2001), a smaller number have tried to assess these impacts for a large number of diverse sites in different country settings (Baland and Platteau, 1996; Shackleton and Campbell, 2001; Ribot, 2001; Mayers and Bass, 1999; Gibson et al, 2000; Enters et al, 2000; Agrawal and Ribot, 2000; Agrawal and Ostrom, 2001). These studies represent important contributions to the comparative study of devolution policies in natural resource management. To varying degrees, they focus on the suitability of different institutional arrangements for collective action (Gibson et al, 2000; Agrawal and Ostrom, 2001), the possibilities for local participation in decision-making within the formal governmental decision-making structure (Ribot, 2001; Agrawal and Ribot, 2000), or the effect of devolution on government objectives for forest production and protection (Mayers and Bass, 1999). While these are important areas of analysis, we believe most take for granted both the 'public goods' interest of the state in forests and the legitimacy of government-sponsored devolution arrangements. We have tried, instead, to focus on local political and livelihood interests – especially of women, ethnic minorities and the very poor – and to consider alternatives to the formal arrangements for devolution that were put in place by governments. From this analytical starting point, we treat rights of local disadvantaged groups as primary, arguing for policy reforms that protect these local rights, while making incremental gains in protecting the public interest, rather than the reverse.

In other words, rather than assess how communities can be instruments for achieving state forest management objectives, we assessed if devolution policies could be a means of promoting rural people's self-determination and economic advancement in forest management, or what we called 'space for local forest management'. We looked at this space in terms of the extent to which local people, especially disadvantaged groups, exercise control over:

- changes in the extent and quality of forest;
- their economic assets and livelihood strategies; and
- decision-making processes related to forest management.

We purposefully use the term *control* to distinguish our work from models of public participation that solicit input to decisions controlled by forest agencies (see Daniels and Walker, 1999, and Rossi, 1997, for discussions of various forms of public participation). Control implies that local people make decisions about forest management themselves, or those who do are accountable to them (Ribot, 1998). We suggest that there may be a need to apply such a framework to better understand the impacts of devolution not only for forest management, but for other natural resources as well.

In taking this perspective, we seek to promote democratic decision opportunities for people who depend on natural resources with 'public good' qualities, such as forests, water and wildlife. Strong arguments exist for recognizing the benefits of these resources to those outside the local community. Watershed management, biodiversity conservation and carbon storage, for example, have effects well beyond the community level (Enters and Anderson, 1998). Yet, historically, many natural resources have so frequently been treated as the domain of the 'public interest' that the fundamental rights of the people living in, or using, them have been overlooked or traded in the name of the greater social good. Even where the rhetoric of poverty alleviation exists, such objectives have been framed in terms of meeting national objectives related to economic development, not in terms of protecting an individual's or community's right to economic self-determination. People living in forest areas, in particular, have been expected to cope with sometimes drastic limitations on their choices and to yield rights of self-determination commonly enjoyed by others living outside of forests (Sundar, 2000b; Scott, 1998; Peluso, 1992; Hecht and Cockburn, 1989; McCarthy, 2000a; Klooster, 2000). Forests and many other natural resources clearly require a multi-stakeholder approach to management; but there has been a lack of 'space' for the poorest users to influence management decisions (Wollenberg et al, 2001). Opening up this space is now the key challenge for natural resource policy.

Our research was not intended to represent the whole story of forest devolution's impacts on local people. Instead, we have tried to elaborate on the issues that have not received adequate attention in assessments and that we felt should be brought into discussions about forest policy reform. To highlight these issues, we focused on the three countries with the longest experience with devolution and its large-scale implementation across diverse contexts. In sharing the lessons from these countries, we acknowledge that considerable progress has been made in some places for some local people; here, we focus intentionally on where *further* progress can be made to foster learning for the next generation of policy development.

The remainder of this chapter outlines how we conducted our research and introduces some of our findings. Following the introduction, case study material from China, India and the Philippines is presented and analysed in Chapters 2 to 4. Chapter 5 then provides a summary of the impacts across all three countries and an analysis of the divergent interests of government and local groups to explain the inherent limitations of devolution as it is currently conceived. We end by identifying where further policy reforms are needed in Chapter 6.

Our findings indicate that devolution policies had a negative impact on the lives of local forest users in many of our case study sites. This was especially true of politically disadvantaged groups who were often unaware of the implications of policy reform or unable to affect policy implementation to protect their interests. While we intentionally looked at cases where policy implementation was likely to be a problem, the results are consistent enough, and their implications serious enough, to suggest that support for the current array of devolution policies should be reconsidered, especially in India and the Philippines. Where alternative strategies for promoting local forest management are available – such as existing, self-initiated local management systems in India or farmer shareholding systems in China – devolution policies should be reformed to make space for these alternatives.

For example, we found that forest cover had increased following the introduction of devolution policies in each of our case countries. This was not true in all locations, as existing local protection systems were sometimes displaced by less effective state-sponsored 'local' institutions, especially in India. A more common problem, however, was that an increase in forest cover translated into a decrease in the accessibility, if not the sheer quantity, of forest resources that were important to local forest users. In many of our sites in India and the Philippines, timber and agroforestry species favoured by governments dominated reforestation and afforestation efforts. These species displaced species valued by local villagers, including those valued for the protection of soils and water sources, medicinal plants, fodders sources, forest products used in local construction and artisan production, and wild foods. Plantations were also established on common lands that had been used for grazing or as agricultural land reserves, devastating the livelihoods of many poor men and women. In China, increases in forest cover better reflected the interests of the poor, as farmers were allowed to choose fruit trees and other species that would yield marketable products. Even in China, however, regulations limited the options villagers had regarding timber species. Where local governments were not held accountable to villagers, plantations were sometimes established on lands used by the poor for other purposes. Thus, increases in forest cover often meant decreases in the availability of forest resources needed by disadvantaged groups.

Our case studies indicate that devolution policies can have a mixed impact on local livelihoods and on other material benefits associated with forest management. Devolution policies provided indirect benefits of legitimacy and visibility to local users in all three countries. Local users enjoyed higher status with such recognition, and could attract resources from the state, donors and non-governmental organizations (NGOs) for their own forest management activities. In some cases, however, devolution policies undermined the effectiveness of existing local forest management institutions by establishing parallel state-sponsored institutions. This was especially common in India and, to a lesser extent, the Philippines. Devolution policies also destroyed as often as they created income-earning opportunities. In some locations, particularly in China, villagers took advantage of devolution policies to grow and market forest products for their own accounts. In India, however, the state often reneged on promises of income-sharing after locals had contributed substantial amounts of labour to forest protection – foregoing other important economic activities associated with forest management. Where policies expanded opportunities for locals to sell forest products directly, poor and minority men and women often lost their place in the trade to elites within, and outside of, the local community. Similarly, access to subsistence products increased in areas where the state had, before, been able to impose tight restrictions on forest access. In contrast, where locals had managed forests *de facto*, contracts with the state, organized under devolution policies, actually decreased local access to subsistence products. In the limiting cases, devolution policies had a negative impact on local livelihoods.

Our studies further show that devolution policies, despite the stated intention of policy-makers, have not always transferred significant decision-making authority to local forest users. In China, far-reaching state control of planning processes was rolled back in favour of planning by households and communities. Villagers in most locations were generally able to make decisions about what to plant, where and when. Taxes and regulations on harvesting and marketing timber, however, remained important tools of state control. More-over, informal pressures on decision-making cannot be ignored in the more remote areas of China, such as Gengma County. Here, villagers came under intense pressure to accept local government decisions that they strongly opposed. In India and the Philippines, the situation was often more discour-aging. Forest departments maintained control over management decisions by controlling work plans, budgets, market outlets and local organizations as strategic points of intervention. In many places, this represented a loss of local decision-making authority. Local organizations – whether self-initiated forest protection groups, *van panchayats* or community-based organizations – that once made many management decisions on their own were subject to the oversight and discipline of the state. Ironically, devolution had simply allowed the state to extend its influence by reducing the state's management costs by 'outsourcing' forest protection and other management responsibilities to various local institutions. Disadvantaged groups often suffered the worst in such outsourcing strategies, as the state formed alliances with local elites to ensure its control over local decision-making.

Our findings also suggest three important points of leverage for expanding the space for decision-making by disadvantaged groups. Firstly, the conceptual frameworks of states[3] and communities[4] often diverge sharply with respect to natural resource management. Efforts to develop a shared framework about the aims of devolution and the mode of its implementation should lead to a more just balance of state and community interests. Research focused on social learning (Röling and Wagemakers, 1998; Maarleveld and Dangbégnon, 1999; Steins and Edwards, 1999) mediation (Blauert and Zadek, 1998), and pluralism (Rescher, 1993; Kekes, 1993; Anderson, et al, 1999) have identified important strategies for facilitating communication among diverse interests in order to define a shared framework in which to make resource management decisions. We take the position that conflict among interest groups will never be elim-inated, but can be better managed than it is now by openly acknowledging differences in interests, by developing institutional mechanisms that coord-inate action, and by incorporating a periodic review of any agreements reached in order to account for changing social and environmental circumstances (Daniels and Walker, 1999; Chandrasekharan, 1997; Fox et al, 1997). We note that actual contact between the central government and local resource users rarely lives up to these standards. There are many constraints to improving dialogue between the central government and disadvantaged groups, beginning with the political structures and cultures that shape policy-making today (Blauert and Zadek, 1998; Mayers and Bass, 1999). We must address the potential for weaker groups to be silenced in any discussions over a shared framework for decision-making about natural resources (Sarin, 1998; Edmunds and Wollenberg, 2001; Baviskar, 2001). Nevertheless, negotiation represents an important mechanism for improving devolution policies and natural resource management generally.

Secondly, strong local capacities for cooperative action can serve to protect and promote local decision-making space. These capacities are sometimes treated as the product of social capital: 'features of social organization such as networks, norms and social trust that facilitate coordination and cooperation for mutual benefit' (Putnam, 1995; see also Coleman, 1990). We take the position that social capital can be nurtured within communities by the state or other actors (Harriss and de Renzio, 1997) through legal literacy and similar programmes. We also argue that such capital can be 'scaled up' beyond the local level (Bebbington, 1998b), and can cross the 'public–private divide' (Evans, 1996) as a powerful force for promoting the interests of disadvantaged groups.[5] We look carefully, however, at the risks of scaling up through federations of forest user groups (see also Shrestha and Britt, 1997). The capacity for cooperative action can be co-opted by the local elite, dominant local groups or outside parties at the local scale, and the risk remains as resource users build regional or national alliances (Harriss and de Renzio, 1997; Jenkins and Goetz, 1999). We also found that individuals did not always want to act on the decision-making opportunities presented to them, but sometimes preferred to yield that right to another group (see Melucci, 1996, for a discussion of the right *not* to participate in decision-making).

Thirdly, the structure of decision-making has a strong impact on equity outcomes. We are concerned here not only with equity between local communities and 'outsiders', but also between different resource-using communities and within single communities. Disadvantaged groups, by definition, suffer under historical inequalities within the larger nation states, and are often treated as second-class citizens, and even as non-citizens.[6] There are also important inequalities between resource-using communities. Those who initiate natural resource protection early can leave neighbouring communities with limited access to resources; lowland communities can impose use restrictions on upland communities in the name of ecological services without paying compensation; and powerful ethnic groups and castes can assert control over resources at the expense of weaker groups. Finally, it is well documented that 'communities' are differentiated by class, caste, gender, race and many other factors (Fortmann and Bruce, 1988; Li, 1996; Sarin, 1998; Rocheleau and Ross, 1995). Inequalities arise not only in terms of access to and the distribution of benefits from natural resources, but in the role that disadvantaged groups play in decision-making processes.

Structures that directly involve and empower disadvantaged groups are more successful than those that allocate control to higher levels of social organization – even, in some cases, including such aggregates as 'communities'. Problems of representation and accountability inevitably crop up when disadvantaged groups are involved only indirectly in making decisions (Wollenberg et al, 2001). Processes should be in place that enable disadvantaged groups to mobilize and negotiate directly with others when they wish to do so. Where indirect representation is necessary, mechanisms must ensure accountability to disadvantaged groups (Ribot, 1999). Ideally, these mechanisms will be flexible enough to account for variations in social and environmental contexts, and changes over time that shape power relations in specific locations (Lindsay, 1998). In some cases, this may mean giving special protection and even preferential authority to those traditionally lacking in power.

In reporting these findings, we must stress that there are no easy or obvious solutions; rather, there are instructive examples of local-level innovation: gains for some that created losses for others, as well as outcomes shaped fundamentally by geographic and historical context. Nevertheless, some patterns do emerge in how devolution policies have affected the decision-making space of the most disadvantaged resource users. We conclude with suggestions for moving policy debates and the practice of natural resource management in new and better directions.

## ORIGINS OF THE BOOK

The impetus for this book grew out of a research project coordinated by the Center for International Forestry Research (CIFOR) from 1998 to 2000. In 1997 the International Fund for Agricultural Development (IFAD) asked CIFOR to coordinate research in Asia to assess devolution policy impacts and their causes. As poverty alleviation is central to the mandates of both CIFOR and IFAD, the impacts of policy on the poorest forest users were of particular concern. We selected India, China and the Philippines as focal countries for the study because of:

- their early efforts to develop forest devolution policies and their role as models for policy development in other countries in Asia and beyond;
- their large populations (a significant portion of which are poor and make their living from forests);
- the large proportion of forests in each country that have been controlled by the state, rather than by individuals or private companies; and
- the differences in the way that the state and civil society interacted in each country.

For each of these countries, CIFOR contacted local government, university and NGO institutions to assemble country-based research teams. The country teams included researchers with long-term involvement in devolution issues and strong field experience.[7] In 1998, the coordinators of each country team then developed joint research questions and hypotheses to be tested across the three countries. Members of the coordinating group met four times during the three years of the project to further refine propositions, develop common methods and share findings. It was during one of these meetings that we reached the conclusion that our findings warranted a book.[8] We wanted to show, from the perspective of poor forest users in several country contexts, how starkly the rhetoric of devolution objectives has diverged from reality. Thus, in August 2000, the authors of this volume assembled to integrate our findings and collectively produce the material presented here.

## RESEARCH FRAMEWORK

The three study teams conducted their work within the same broad research framework. This framework consisted of three common research questions, a shared understanding of what we meant by devolution and the kinds of policies in which we were interested, and a common definition of what we meant by decision-making space. At the same time, study teams developed their own, largely qualitative, research methods. They also highlighted different aspects of the research framework, depending on local circumstances. We believe that instructive comparisons can be made among the three case studies, and have tried to draw conclusions from such comparisons in the final chapters. However, there are enough differences in how the research was conducted among the case studies that most of our comparisons are at a fairly general level. Specific conclusions and recommendations are made only for the countries in which the research was conducted.

### The research questions

To reach our ultimate goal of assessing devolution policies in terms of their impact on the decision-making space of local forest users, we realized that we needed empirical information along three lines of enquiry:

1   How have devolution policies affected local people's space for making decisions about forest resources, livelihood choices and local governance?
2   Why have these impacts occurred?
3   How have local people sought to protect their ability to make important decisions, with or against devolution policies?

In answering these three questions, we aimed not to produce a comprehensive evaluation of the impacts of a country's policies, but rather a well-substantiated picture of policy impacts across a range of types of devolution and sites, with an in-depth understanding of cause and effect at each site.

The first question led us to understand the degree to which devolution policies enabled or encroached upon the decision-making space of local forest managers. We identified key devolution policies in each country, described below, and then assessed these policies' impacts on the ability of people living near forests in our study sites to influence forest management decisions, improve people's livelihoods and protect or enhance the forest resources that people valued. We also looked at more conventional 'forest agency' indicators, related to forest cover as a point of reference, and because we wanted to point out where official assessments diverged from those of local users.

We then needed to be able to explain why these devolution impacts occurred. Why did impacts often diverge from the stated intent of the policy and why did they vary from site to site? One of the key assumptions underlying the research was that policy-making is not a linear 'top-to-bottom' process. We needed to take into account the actions and responses of local people, advocacy groups, other private-sector interests, different levels and sectors of government, and local ecological conditions to explain policy outcomes (Mayers and Bass, 1999).

We therefore examined four conditions that we thought, from past studies and our own experience, most strongly influenced policy impacts. We expected that:

1   the extent of local people's economic or cultural dependence on the forest would influence their interest in maintaining it;[9]
2   the strength of social capital among forest users, and between them and other groups, would influence their capacity to maintain the forest and assert their interests in it;[10]
3   the type and strength of devolution policy in the community, and possible contradictions with other policies, would influence the policy impacts;[11] and
4   external actors and institutions could play a role in supporting or derailing devolution policies.[12]

Sites in each country were selected to demonstrate variability in these conditions, although the research teams did not always find all four conditions salient in each site.

The third question enabled us to explore the processes by which local people were able to protect, maintain or expand their decision-making space parallel to, or in spite of, devolution policies. We looked at cases of local forest management that pre-dated the present era of devolution policies, beginning in the early 1980s. We examined self-initiated forest management in Orissa, the *van panchayats* of Uttar Pradesh (now Uttaranachal), India; and indigenous management among minorities in Yunnan, China; and in all three study regions of the Philippines. We explored how these existing systems sometimes were co-opted or came into conflict with state-led policies. We examined the impacts of these conflicts and local people's responses. We also gave special attention to how people organized to change devolution policies through federations of forest users, alliances with NGOs and others, and we examined historical and contemporary social movements for tribal self-rule. These efforts at modifying or overturning devolution policies were often the starkest indications of the need for more democratic forms of decision-making that allow local people to exercise greater control over forest management practices and decision-making processes.

## Defining devolution

We defined devolution as the transfer of authority over forest management decision-making from central government bureaucracies to local civil society actors, generally forest users and user organizations not created or controlled by government (Fisher, 1999). Authority is assumed to include decisions about what counts as forest and how forest benefits are used. We include in this definition of devolution a transfer of authority that others might call 'privatization': the transfer from collective management institutions to individual households in China. In contrast, we call the transfer of decision-making authority to local governments 'decentralization'. Decentralization *per se* was not, however, the focus of our study.

We developed a typology of devolution policies that helped us to identify the range of policies in practice. We classified devolution policies according to three characteristics:

1   the actors involved in the transfer of authority;
2   the set of rights and responsibilities devolved; and
3   the relationship to existing local management practices and decision-making processes, especially those of the poorest forest users.

We defined policy as formal government initiatives, whether in the form of administrative orders, legislation or court rulings, though we recognize that policy as it is enacted 'on the ground' is produced by a wide array of actors through many informal processes.

The first dimension described the actors involved in the transfer of forest management authority. We looked at the three trends most common in Asia:

• from central and local government to village-level civil society groups;
• from central government to local government;
• from collective institutions to households or individuals.

The borders among any of these actor categories can be blurred. Joint Forest Management (JFM) committees in India and Community-based Forest Management (CBFM) organizations in the Philippines might be seen as either civil society actors or creatures of the government bureaucracy, depending upon how they function. Similarly, there is some doubt about whether village committees (VCs) should be understood as local governments or as arms of the central government. We try to make this distinction clear in the case study chapters by focusing on how local organizations were created and to whom they are accountable (see Ribot, 1998, and Agrawal and Ribot, 2000, for a discussion of accountability).

The second dimension described the rights and responsibilities transferred. We focused on three sets of rights and responsibilities:

- managing the forest as a whole;
- forest products; and
- governance of administrative units.

The focus of debates about devolution and forestry has often been about transfers of responsibilities or rights to manage forests. Yet, devolution has also taken place with respect to specific forest products, as in India with the transfer of rights to harvest and market some non-timber forest products (NTFPs) from state enterprises to individual farmers. New forms of political governance can also transfer responsibilities or rights for forest management among actors. The Local Government Code in the Philippines, though not specifically focused on forest resources, gave local governments more authority to make decisions about forest management in their jurisdictions.

In fact, as our case studies indicate, local civil society actors and local governments never exercise unfettered control over all aspects of forest management (see also Lindsay, 1998). It is therefore necessary to break down forest management 'authority' into specific rights and responsibilities (Fortmann and Bruce, 1988). Rights to choose species and determine where and when to plant them have been devolved from the Forestry Department to households in China, for example; but households have limited rights to harvest or market many valuable forest products, including timber. It is also important to distinguish between rights and responsibilities, as they have not always been devolved together (see Enters et al, 2000). CBFM programmes in the Philippines, for example, provide villagers with few rights to choose species and dispose of trees, but 'contract' villagers to be responsible for forest management. JFM also requires villagers to protect forests for a share of benefits, a share largely determined by the Forest Department from a forest shaped primarily by department priorities. Such partial devolution creates conflicting incentives for forest management and limits the space for meaningful decision-making at the local level.

We also note that there is often a large gap between paper authority and the ability to make meaningful decisions. In part, this is due to the large influence that foresters and other officials all along the bureaucratic hierarchy can exercise over the implementation of policy (Gupta, 1995; Mayers and Bass, 1999; Malla, 2001; Magno, 2001), and the many pressures that are brought to bear on them that make implementation politically charged (Hirschmann, 1999). It is also an issue of creating the capacity to act on a legal right. Decision-making authority, in our view, includes not only the legal right to make decisions, but the resources, political support, technical and institutional conditions that allow local people to make meaningful decisions (Fisher, 1999; Agrawal and Ribot, 2000; Manor, 1999).

The final dimension related devolution policies to the different local contexts in which they were to be implemented. We looked at existing forms of management and decision-making by the poorest forest users in several of our sites.[13] We characterized the relationship between devolution interventions and existing local practices as:

- conflicting;
- mutually supportive; or
- neutral: no obvious interrelationship.

Devolution in Asia is usually treated as a process by which governments determine the nature and scope of the responsibilities and rights transferred. These processes, however, often occur in the context of existing management systems where local actors have already created space for themselves to make management decisions (see Arnold, 1989; Ostrom, 1999; Mayers and Bass, 1999). Federations of self-initiated community forestry organizations in Orissa, for example, have developed their own decision-making processes and priorities for forest management, disrupted by the introduction of JFM in the state. CBFM programmes in the Philippines were also laid over existing forest management institutions at the village level, and imposed new management guidelines and decision-making procedures upon these institutions. In fact, the tendency of central government officials to ignore the existence of these local practices is a major reason for conflict over devolution and other forest policies (Gilmore and Fisher, 1997; Anderson et al, 1999). Examples of supportive policies are harder to come by. Perhaps the Indigenous Peoples Rights Act (IPRA) in the Philippines, and the Chinese government's tolerance of Miao traditional leadership's role in forest management, are the best examples we have found thus far – though IPRA is still too new to assess, and the rights of national minorities to govern forests in China are tenuous. Our research, therefore, gave particular attention to struggles where government initiatives appeared to be squeezing out or co-opting local initiatives, but where local resistance to such a process was strong.

These categories are not hermetically sealed. Local initiatives were sometimes in response to, and influenced by, actions of the central state and vice versa. In China, policies in support of forest shareholding agreements evolved, in part, out of local initiatives to overcome problems of fragmented forest holdings. There is now a struggle to determine what kinds of shareholding contracts will be permitted and supported by the state. Federations of community forestry organizations in India obtained recognition and support from the central state and donors, and determining who is influencing whom can be complicated. Nevertheless, we believe that it is crucial to make visible the local management and institutional context – pre-policy – if we are to accurately assess what impacts devolution policies have had on local decision-making space.

With these classifications in mind, we chose cases that reflected an array of different types of devolution. In India we examined:

- contemporary joint forest management (JFM) in Orissa and Madhya Pradesh that theoretically transfers management rights from the state to community-based organizations; we include sites where such policies conflict with efforts by civil society groups to develop community-based forest management (CBFM);
- contemporary non-timber forest product (NTFP) policies in Orissa and Madhya Pradesh, where nationalized control over harvesting and marketing has been devolved to individuals;
- movements for tribal self-rule and Provisions of the Panchayats (Extension to Scheduled Areas) (PESA) Act 1996 as examples of devolution that take place through local governance institutions, especially those of indigenous groups;
- *van panchayats* as examples of existing CBFM practices and institutions that conflict with the JFM policies of the Forest Department.

In China, we looked at:

- household management systems where rights and responsibilities for forest management were transferred from collectives to individual households, though rights over some forest products, especially timber, were not;
- collective management systems where central government reigns over collective forest management were significantly loosened, and the decision-making of collectives was, in some cases, accountable to villagers;
- shareholder systems that devolved forest benefits from the state and/or collectives to households while retaining for the state most forest management decisions;
- self-initiated shareholder systems that reflect farmers' own initiatives to overcome problems in exercising their legal control over forests, supported by policies that allow leasing and contracting of live stands;
- local management in protected areas where limited forest use rights have been granted to villagers by the state in exchange for protection services;
- traditional management systems of the national minorities as examples of local initiatives that were influenced by national policies.

In the Philippines, the differences among cases were often more in name than in actual practice:

- Integrated social forestry (ISF) programmes, where communities are given either communal leases (Community Forest Stewardship Agreements) or individual household leases (Certificate of Stewardship Contracts) by the state. Communities are limited to agroforestry farming under the leases.
- Community-based forest management of 1995, where communities are given communal leases that are good for 25 years but renewable for another 25 years. Communities are involved in agroforestry, reforestation, timber stand improvement of residual forests and assisted natural regeneration of regenerating stands, which provide greater benefits but few decision-making opportunities.

- Ancestral domain claims where indigenous cultural communities are awarded tenurial rights by the state to ancestral lands in the form of a Certificate of Ancestral Domain Claims (CADC). Again, communities are responsible for forest management as specified in lease agreements drawn up by the Department of Environment and Natural Resources (DENR).
- Indigenous Peoples Rights Act (IPRA) – this differs from all the other devolution policies by granting extensive tenure rights to indigenous cultural communities and recognizing indigenous customary law as a legitimate decision-making process for forests.
- Local Government Code – the authority for forest management was devolved from the DENR, at the national level, to local governments.
- Community-based forestry initiated by people's organizations (POs) and NGOs – this involved local people forming their own institutions and management guidelines, which now confront efforts by the state to bring them into the fold of CBFM.

Although we refer to the specific policies studied as devolution policies, we stress that this does not imply that the policies necessarily transferred authority to local civil society actors or local governments. Rather, we describe policies in this way because of their apparent objectives. We use a different vocabulary (decision-making space) in assessing whether a transfer of meaningful authority really occurred (see the following section for a discussion of decision-making space).

We also recognize that devolution is but one means of creating space for communities in forest management. In some cases, communities may decide that the best way to create space for themselves is by making new demands on state services, such as for technical support or funding (Evans, 1996; Blauert and Zadek, 1998). This may imply a stronger role for central governments, rather than government retreat. What is crucial in such circumstances is that communities have the power to help bring about a fair distribution of rights and responsibilities between themselves and governments. Restructuring government–civil society relations does not necessarily mean minimalist government. But it does require government that is accountable to local interests and that balances these interests justly when making decisions about public goods. Our research noted that local forest users sought specific state services as a means of supporting local decision-making.

## Defining space for local forest management

Devolution can be thought of as a process that shifts the decision-making space related to local forest management from centralized government to local communities or local government. This space is multidimensional. To observe and report about control over decision-making, we defined this space in terms of three dimensions: the ability to control decisions about the extent and quality of forest resources; livelihoods and income; and political processes related to forest management.

We sought to determine the extent to which *locally meaningful* spaces were created for different local forest users, particularly the poorest users. This meant drawing on the conceptual frameworks of these users as much as we could. Our criteria for evaluating the impact of devolution policies on local decision-making space reflect our understanding of what local forest users consider meaningful, based on past field experience and readings.[14] It also meant that we needed to understand the trajectories of change in that decision-making space for a given historical context and place. We wanted to recognize the different starting points of forest users and communities and to acknowledge where progress had been made under difficult circumstances. Studies of self-initiated forest management, for example, were included to compare the types of decision-making spaces that these groups enjoyed prior, or parallel to, those created under devolution policies. Villagers in Uttar Pradesh and Orissa, for example, did not interpret the extension of JFM to village forests and common lands as devolution at all, arguing that the Forest Department had no claims on these forests, in the first place. Women in many JFM programmes in India observed that there had been practically no change in their rights to decision-making and that, in fact, their customary forest access had declined rather than increased. Similarly, whether or not the Local Government Code in the Philippines created space for local decision-making varied from the perspective of dominant or minority ethnic groups. Evaluations of devolution policies varied over time, as well. A move from large-scale collective to smaller-scale collective management in China may have been warmly welcomed by villagers ten years ago, but appeared inadequate to meet the new market opportunities available today. We therefore account for multiple and changing perspectives as we interpret and analyse different devolution policies.

To assess the impacts of devolution policies for different groups of people within and across villages, we encouraged our study teams to assess each of the dimensions of decision-making control across ethnic groups, gender and the non-elite/poor and elite/better off.[15] Where appropriate, forest quality and forest cover were assessed to clarify debates about the impact of devolution on the 'public goods' associated with forests (Guha, 2001).

Table 1.1 provides a summary of the key variables that we developed as potential indicators of the impacts of devolution. These variables were developed during the course of our research and therefore are not systematically covered in these case studies; rather, they reflect dialogue between our study teams and local forest users, NGO staff, local government officials and others outside of the central forest departments. One step towards developing a shared framework of devolution between local and state groups is to help integrate these locally based measures of desirable impacts into formal assessments of community-based management programmes that hitherto have relied heavily on government department perspectives.[16] We recognize that the definition of space itself will change over time for different users and communities, and this is something we suggest could be monitored to keep frameworks for devolution dynamic and relevant.

**Table 1.1** *Variables defining the spaces for local forest management*[17]

| Dimensions of Space | Indicators | Equity Impacts[a] |
|---|---|---|
| **Forest quality** | Control over forest resource extent and quality<br><br>Forest cover | Example: CBFM programmes in the Philippines encourage agroforestry species that produce building poles used by men at the expense of fodder grasses used by poor women. |
| **Livelihoods and well-being** | Access and control over forest land, forests and products<br><br>Access and control over markets, credit and transport<br><br>Level and control over income<br><br>Other well-being indicators<br><br>Security | Example: Farmers who are poorly connected politically, and who are isolated and poorly educated cannot take advantage of forest stand leasing in China. |
| **Political decision-making and governance** | Over forest management[b] and policy decisions[c]<br><br>Access to technical information, resources and personnel<br><br>Control over local institutions and organizations<br><br>Control over relations with others inside and outside the community[d] | Example: Women from lower castes in India have been replaced by higher caste men in the leadership positions of user groups that they initiated. |

*Notes:*  **a**  Who is affected (by gender, ethnicity and elite/non-elite status), the quality of the impact on each group, and the trend in change for each group.

      **b**  Management activities that forest users may wish to exert control over included:
- decisions about land-use regulation;
- choice of species;
- technical methods/silviculture;
- forest maintenance;
- harvesting;
- sales;
- use of income.

      **c**  Policy features that forest users may wish to exert control over included:
- formal right or informal ability to interpret policy at the local level;
- rights or abilities to implement policy at the local level (including enforcement);
- rights or abilities to reject policy at the local level;

- representation in government decision-making processes;
- channels of communication with government officials;
- access to policy information.

d   Relations of importance to forest users that we considered included:
  - the right or ability to protest;
  - access to and influence over NGOs, media and donors;
  - control over the representation of community interests by others (including researchers);
  - the right or ability to form alliances with other forest users at larger geographic scales.

We worked with the assumption that the more control local people have over these three areas of local forest management, the more they will be able to negotiate effectively with the state and other parties about fair trade-offs in local and public interests. Thus, in measuring the extent of control over decision-making, we are not advocating that any group adopts full control over a forest, but rather that they gain more capacity to exercise their interests in competition with other more powerful groups.

## Methods

The research teams made extensive use of qualitative research methods, especially standard qualitative interviews (Nachmias and Nachmias, 1987) and the group interview techniques associated with participatory rural appraisal (Chambers, 1992; Jackson and Ingles, 1998). Archival work was also important, especially in establishing an historical context for recent devolution policies, identifying specific language in policy documents, and in addressing government assessments of forestry and devolution. Participant observation was important in each country, as well – in India through the team leaders, who are themselves active in forest management organizations that work with forest user communities; in the Philippines through partnerships with local government officials and activists in community forestry; and in China through partnerships with local forestry officials involved in implementing devolution policies.

As discussed above, specific research sites were chosen that varied according to the extent of local people's economic or cultural dependence on the forest; the strength of social capital among forest users and with other groups; the type and strength of devolution policy in the community and possible contradictions with other policies; and relations with external actors and institutions. The resulting cases consequently demonstrate considerable variation in the implementation of devolution policies. The finding from these cases are synthesized in the three country case study chapters that follow.

# 2

# The Promises and Limitations of Devolution and Local Forest Management in China

*Liu Dachang and David Edmunds*

## INTRODUCTION

During the early 1980s, China began experimenting with devolution policies for local forest management. These policies included the transfer of forest management from collectives to households, the involvement of villagers in the management of state forests, and the shift of some decision-making authority in forest management from the commune and forestry department to more local entities, such as village committees (VCs).[1] In this chapter, we examine the impacts of these policies, explain why they have not always lived up to the expectations of either farmers or government officials, and draw some general conclusions about the devolution experience in China.

China is an interesting place to study devolution for at least four reasons:

1   China is a strong state with a reputation for fast and effective implementation of policy on a grand scale. Local forest management doesn't happen by accident here: either the state actively encourages it for its own reasons, or farmers devise ways of pressuring the state to allow the exercise of local authority. Getting the policy right is more crucial than in places where policies are in no danger of being implemented, or where they can be ignored with impunity.
2   China has relatively few non-governmental organizations (NGOs), independent research institutes or donors that could mediate between the state and farmers. The state and farmers usually confront one another directly, without such intermediaries. As such, there is a premium on building good relationships across the public–private divide, something that other countries struggle to achieve even with the help of mediators (Evans, 1996).
3   In China, devolution often involves a transfer of authority from collectives to individual households, as well as from the state to village committees. The Chinese experience provides guidance on whether and how to allocate forest management to households.

4   Chinese officials were trying to build local social capital and institutional capacity by collectivizing rural production, including production of forest products. Comparing the kinds of institutions created by the state with the kinds that have emerged from local initiatives or localization of state policies provides insights into the most effective division of roles among states and farmers in forest management.

Our study found that, in general, devolution policies have expanded local decision-making authority in forest management, increased the material and environmental benefits that farmers receive from forests and improved forest cover. Progress has been greatest where policy has been adapted and effectively localized, where local social capital has been strong, including good relationships built between farmers and local officials, and where the state has helped to improve local technical expertise and marketing conditions. Still greater successes could be achieved by reducing the regulation of forest-product harvesting and marketing, and by improving incentives to farmers for managing forests well. Self-initiated shareholding systems may be one of the best ways forward, though government will have to ensure that social inequality does not increase greatly under such a programme.

## Devolution and Local Forest Management in Context

In rural China, especially in the 12 southern provinces (Anhui, Fujian, Guangdong, Guangxi, Guizhou, Hainan, Hubei, Hunan, Jiangxi, Sichuan, Yunnan and Zhejiang), rural communities and households live in and around forests and are highly forest-dependent. They harvest timber, fuelwood and non-wood forest products, such as nuts, fruits, wild vegetables, bamboo shoots, fungi and mushrooms, medicinal herbs, spices and fiber. These forest products are for on-farm consumption and market sale, or for the exchange of manufacturing goods. The forest provides important grazing land for many farmers with domestic animals. Forests also provide environmental services and are often part of sustainable agro-ecological systems in these provinces. Ethnic minorities live in most of these areas, and forests can have religious and cultural value for them, as the persistence of holy forests in some areas would suggest. Han communities associate forests with historical events, social gatherings and cultural practices, as well. Hence, forests are quite important in people's lives in south-western China.

China's forests can be grouped into two broad tenure categories: state forest and non-state forest. State forests are those that fall directly under the supervision of the central government's Ministry of Forestry (MoF). Non-state forests have fallen under the direct supervision of other institutions, such as the collective or, more recently, the administrative village and even the household – though, as we will see, the central state still exerts considerable control over their management. Non-state forest represents about 65 per cent of the total forest area in Yunnan and Sichuan provinces. In the other ten southern provinces, the proportion of non-state forests is as high as 90 per cent of the forest area (MoF 1995, pp85–105). Local communities and households not only enjoy significant tenure and management rights in non-state forests, but are also involved in the management of parts of the state forest in these provinces, especially in Sichuan and Yunnan.

## Non-state forests

Over the agricultural collectivization era (the mid 1950s to the early 1980s), all forests other than state forests were called 'collective forests' because they were managed by the collective (either a production brigade or production team[2]), and the collective had far-reaching rights to dispose of the trees as it saw fit, though individuals enjoyed rights over some trees near homesteads at different times over the period (Liu, D, 2001).

During the early 1980s, following reforms in the agricultural sector, and modelled on them, China initiated nation-wide reforms in collective forest management. The early forest reforms transferred responsibilities and rights for forest management from the collective to individual farmer households, and from the commune and forestry department to the production brigade (now the administrative village). These reforms also distributed use rights to wastelands – lands without trees but considered by the forestry department to be suited to forestry – to farmer households. Those who planted trees on wastelands sometimes enjoyed extensive tenure rights to them. During the late 1980s and early 1990s, further reforms legalized the leasing of use rights to forest land by administrative villages and households, even permitting trade in live trees and stands. This description does not quite capture, however, the diversity of tenure arrangements that grew out of the reforms. Administrative villages often retained substantial decision-making authority, or tried to, regarding when trees were cut and what was done with the receipts, even on wastelands. Farmer households sometimes acted in their individual interests and sometimes formed shareholding groups or other arrangements for sharing rights and responsibilities. As a result, tenure arrangements and forest management institutions have become quite diverse in China, and it is inappropriate to call them all collective forests. The term 'community forests' is also not appropriate, with its implication of informal decision-making by farmers acting in concert with one another. Many forest management decisions were, instead, the product of struggles between the administrative village, acting largely as an arm of the central government, and individual households acting on their own. Liu (Liu, D, 2001) prefers the more general term 'non-state forests' to refer to this broad and diverse category of tenure and management arrangements, and we use it here, too. Below, we list the forest reforms associated with non-state forests before describing them in some detail:

- *household-based management*:
  - family plots (*ziliushan* in Chinese);
  - responsibility hills (*zerenshan* in Chinese);
  - bamboo forests (in Suining, Hunan);
- *collective management*:
  - modified collective management;
  - shareholding systems;
- *self-initiated shareholding systems*:
  - farmer–farmer collaborations;
  - company–village partnerships (collaboration between outside institutions and an administrative village or an informal organization of farmers);
  - collaboration between outside individual(s) and a village/households;
- *ethnic minority management systems*.

## Household-based management

### Family plots (ziliushan)
Beginning during the late 1970s in a few provinces, such as Yunnan and Guizhou (Yunnan Provincial Department of Forestry, 1987, pp80–81; Compiling Board of Guizhou Annals of Forestry, 1994), and spreading through the early 1980s to the rest of China, collective wastelands and sparsely stocked forest lands were distributed to farmer households to encourage villagers to plant trees to meet their subsistence needs for forest products, especially fuelwood. The collective remained the owner of the land, but villagers could plant species of their choice and dispose of trees that they planted more or less as they wished. The government even legalized the transfer or lease of use rights to other parties during the early 1990s. As we will discuss later, however, taxes, regulations and other factors constrained the types of decisions that farmers could realistically make, and limited the exercise of these new tenure rights.

### Responsibility hills (zerenshan)
During the early 1980s, part or all of the collective forest was contracted to farmer households. Forests that were contracted in this way were still collective property; but households took responsibility for forest management. Accordingly, income from the forest was split between the collective and individual households. The amount of income earned by each household was determined based on the terms of 'contracts' between the collective and households. The terms varied widely from place to place. The objective of the responsibility hill system was to improve the management of existing forests by modifying management methods within the collective system. The rationale was that if farmers were given a share in the benefits of forest management, they might contribute more time and energy to the tasks of reforestation and forest tending and guarding, an assumption drawn from the success in household-based agricultural production – the Agricultural Production Responsibility System (APRS).

Most responsibility hills have evolved in two directions since the 1980s:

1   They were incorporated within family plots in many areas during 1983–1986 (Chen and Gao, 1997; Compiling Board of Hubei Provincial Annals of Forestry, 1989; Compiling Board of Guizhou Provincial Annals of Forestry, 1994). The absorption was not sanctioned by law. Instead, it was a result of initiatives by government officials and farmers who favoured expansion of the household-based management. The process was stopped by central government policy in 1987 (MoF, 1988); but areas already converted to household management were not required to return to responsibility hill systems.
2   Responsibility hills have returned to collective management as the forestry department decided that the system increased deforestation and the department desired greater control over this category of forests. As a result, responsibility hills as a form of management are not nearly as widespread as in the early stage of forestry reform.

### Bamboo forests

This is a special case, found in Suining, Hunan. County and township governments allocated collective forests to farmer households during the forest reform in the early 1980s, then resumed collective management in 1987. In 1997, bamboo forests were again redistributed to households through contracts with a term of 30 years. This kind of tenure arrangement can be viewed as a type of 'responsibility hill', although it differs in when the forests were distributed, the type of contracts developed and the terms of income distribution. Bamboo farmers in Suining enjoyed greater control over management decisions and the distribution of income than farmers in the typical responsibility hill system.

## Collective (village-based) management

There are two forms of collective management. They are included in the discussion of devolution because they represent a shift of decision-making authority, even though the forest management institutions are nominally the same as those of the commune era of centralized management.

### Modified collective management

Even after the introduction of household management, many villages retained part of their forest under collective management. In some villages, as noted, responsibility hills were even returned to the collective. Today, this remains an important form of forest tenure and management.

During the commune era, commune and forestry department officials made all the important decisions about how much area to plant in forest, what species to plant, when to harvest and so on. Leaders of production brigades and production teams did little more than implement decisions made by officials at higher levels of the state bureaucracy. Today, village committees and the leadership of *cunminzu* (the equivalent to a production team; see endnote 2) enjoy expanded political space to make decisions about how forests will be managed. They control collective forests, though timber harvesting is subject to a cutting quota determined by higher levels of government, and any sales of timber must be made to state-owned timber companies at fixed prices. With these caveats in mind, they can decide what kinds of forest products may be harvested and in what quantities, where harvesting may take place, and what species to plant. They can even ignore or override orders from township government and the forest department, though they must do so with caution, as the township government has the power to turn them out of office. However, this reform did not necessarily expand the decision-making authority of villagers. Much depends upon whether villagers can hold the village committee accountable for their decisions, as we discuss in more detail below.

## Shareholding systems

Shareholding systems have evolved in two directions over the last two decades:

1   In some areas, such as Sanming, Fujian Province, collective forests were distributed to farmer households in the form of shares, rather than distributed physically, as in the cases of family plots and responsibility hills.
2   Forests distributed to farmer households during the early 1980s were returned to collective management in the form of a shareholding system promoted by government. The shareholding system in Jinggu, Yunnan, is an example. Under shareholding systems, forests are managed collectively, all residents in a village are entitled to an equal number of shares, and profits from forests are distributed to households according to the number of shares that they hold. Village committees make almost all of the important management decisions, with varying degrees of accountability to farmers.

Shareholding systems still cover relatively little forest area. Government officials and some forestry academics have advocated them as a way of overcoming the disadvantages of household-based management, especially the fragmenting of forest holdings and associated inefficiencies and technical problems. Local households, however, are not enthusiastic about the idea. They have either initiated shareholding systems of their own style (discussed in the following sub-section), or maintained household-based management without collaborating among themselves or with outside institutions.

## Self-initiated shareholding systems

Unlike shareholding systems, self-initiated shareholding systems describe joint or collaborative efforts that are initiated outside of the formal government administration. Collaboration is focused primarily on establishing plantations by pooling land, capital and technology. Farmers may collaborate with each other, but often seek the involvement of the forestry department, timber companies and other outside parties, as well. Income from the partnerships is again distributed according to shares, which are, in turn, allocated according to a party's contributions to management activities. In the case of company–village partnerships, the administrative village or the informal farmers' organization normally contributes land (and labour, in some cases) and the company contributes capital and technology.

There are three differences between the shareholding system and the self-initiated shareholding system. Firstly, the shareholding system was initiated by the central government, while the self-initiated shareholding system is initiated largely by non-governmental actors. Company partnerships with administrative villages represent an ambiguous case, as the administrative village often acts as an arm of the central government. Farmer partnerships with forestry department officials are also ambiguous. We include these examples here, however, as many of the officials involved in such partnerships are pursuing local interests, rather than following directions from those higher in the administrative hierarchy. Secondly, the shareholding system largely concerns the distribution of existing collective forests to households. In contrast, the self-initiated shareholding system is more about the pooling of 'household lands' (lands where use rights are held by households) for plantation establishment. Thirdly, all shareholders of a shareholding system are residents in a given village, while shareholders of a self-initiated shareholding system can be farmers in a village, officials and staff members from the public sector, or other individuals and institutions outside of the community.

This form of management became possible after farmers were free to lease land. Contributions and shares are specified in the contracts, providing farmers, in particular, with a sense of security that their interests will be protected. The self-initiated shareholding system now covers only a small area of forest in China, but is spreading quickly and can be found in all of our research sites.

## Ethnic minority management systems

In China, there is some tolerance for the forest management systems of ethnic minorities. The Miao traditional leader, for example, continues to exert considerable authority over management decisions in parts of Guizhou Province. Until recently, minority groups in Gengma County were allowed to continue swidden systems in a mixed forest/farm landscape. Ironically, recent devolution policies seem to threaten minority management systems that survived the era of centralized management. Minority management systems will be discussed by site as it is difficult to generalize about them.

## State forests

There are two institutions for managing state forests in China. One is management exclusively by government forestry departments and state forest companies, as in north-eastern China. The other is participatory management – where management of a portion of a state forest is contracted or commissioned to local communities and/or households, similar to joint forest management (JFM) arrangements in India. The practice can be seen in many areas of southern China, but is most widespread in Yunnan and Sichuan provinces.

Participatory management of state forests was adopted in recognition that local people are highly forest dependent, and improved forest management cannot be achieved without their active support. Under participatory management, villagers are responsible for protecting forests from fire, tree theft and land clearing for food crops. In return, they have access to certain forest products, particularly fuelwood, poles and medicinal and food plants, and they sometimes exert influence over management decisions, such as species composition and harvesting schedule.

## RESEARCH METHODS

The research in China was carried out in the provinces of Guizhou and Yunnan, in south-west China, and Hunan Province, in central China. Research questions were studied at both county level and village level. Six counties were selected from these three provinces (three in Yunnan, two in Guizhou and one in Hunan) and extensive field data was collected in 15 villages (see Table 2.1), supplemented by limited research in other villages in the counties. In addition, this chapter uses data and examples from work conducted by Dev Nathan and Govind Kelkar in China. Sites were selected to represent diverse types of devolution. We also sought a mix of Han and national minority sites, and sites with different ecological conditions and forest types. These site characteristics are summarized in Table 2.1.

## IMPACTS OF DEVOLUTION

### Impacts on the decision-making space for local communities and farmer households

The decision-making authority of rural communities and households has expanded over the past two decades in China, one of the objectives of the 'sanding' policy during the early 1980s (see Liu, D, 2001). Starting from a point where officials at the commune or higher level made nearly all decisions about how to manage forests, there is, today, much more local decision-making. In general, space was created for independent decision-making by households, shareholders and administrative villages, with respect to managing forests. There are, however, important qualifications to this assessment, which are discussed below.

**Table 2.1** *Characteristics of research sites*

| Sites and location | Current types of devolved forest management | Key villages investigated | Primary ethnic groups in key villages | Distinguishing features | Forest types |
|---|---|---|---|---|---|
| Chuxiong, central Yunnan | Household-based management: <br> • private plots <br> • responsibility hills <br> Collective management: <br> • modified collective management | Pingzhang Qingshan Yici Yunqing | Han, Yi, Miao, Bai | Leasing wastelands[3] to households during 1994–1996 <br> Indigenous forest management of Yi people, including small watershed protection, temple and village forests, as well as rhododendron management | Chinese fir and other coniferous forests, small patches of deciduous forests, especially oak and rhododendron |
| Gengma, south-west Yunnan | Household-based management: <br> • private plots <br> • responsibility hills <br> Collective management: <br> • modified collective management <br> Participation of village/households in management of state forests | Dazhai Mangmei Mangsa | Dai (similar to Thai), Wa, Lahu, Lisu | Distributing shifting cultivation lands to households during the early 1980s <br> Leasing[4] wastelands (shifting cultivation lands) to outside institutions during the 1990s <br> Indigenous forest management, including holy forests and trees | Chinese fir and other coniferous forests, larger patches of deciduous forests, rubber and other subtropical species |
| Jinggu, south Yunnan | Household-based management: <br> • private plots <br> • responsibility hills <br> Collective management: <br> • modified collective management <br> • shareholding system <br> Allocation of state forest to villages/households (direct transfer of ownership) <br> Participation of village/household in state forest management | Wenlang Wenzhao Wenzhu Yixiang | Han, Dai, Yi, Hani,Lahu, Muslin, Bai | Distributing shifting cultivation lands to households during the early 1980s <br> Leasing wastelands during the 1990s <br> The shareholding system was initiated by government during the 1990s. Parts of forests under household management were required to combine with forests under collective management, and villagers were distributed shares and became 'shareholders' | Chinese fir and other coniferous forests, larger patches of deciduous forests, rubber and other subtropical species |

| Sites and location | Current types of devolved forest management | Key villages investigated | Primary ethnic groups in key villages | Distinguishing features | Forest types |
|---|---|---|---|---|---|
| **Jinping, east Guizhou, provincial border area, not far from Suining, Hunan** | Household-based management:<br>• private plots<br>Collective management:<br>• modified collective management (including collective forest farms)<br>• shareholding system<br>Self-initiated shareholding system | Dicha | Dong (influenced by Han culture, the Dong are now similar to Han in life style and agricultural production) | The first county in Guizhou to incorporate all responsibility hills into family plots in 1985<br>Diversified forms of self-initiated shareholding system: farmer–farmer and between villages and outside institutions<br>Provincial pilot area for policy reform of collective forest management | Chinese fir and other patches of deciduous forests |
| **Libo, south Guizhou, provincial border area** | Household-based management:<br>• private plots<br>• responsibility hills<br>Collective management:<br>• modified collective management<br>Self-initiated shareholding system | Datu<br>Shuiwei | Miao, Shui | Villages provide lands for social groups, such as youth league, women's union and schools free of charge.; these organizations established their own plantations on the village's lands<br>Legalization of trading live stands<br>Leasing forest lands to outside individuals<br>Indigenous forest management of Miao | Chinese fir and other coniferous forests, small patches of deciduous forests |
| **Suining, west Hunan, provincial border area, not far from Jinping site** | Household-based management:<br>• private plots<br>• bamboo forests<br>Collective management:<br>• modified collective management<br>Self-initiated shareholding system | Fujiawan<br>Tiantang | | Reversing responsibility hills in 1987 to collective management, which is now the dominant form of forest management<br>Distributing bamboo forests to households in 1997<br>Leasing forest lands to outside institutions | Chinese fir and other coniferous forests, small patches of deciduous forests<br>Significant bamboo forests |
| Total | | 15 | | | |

Source: compiled based on data collected by the field teams

## Household-based management

Family plots provide farmers with more direct control over forest management decisions than the various forms of collective management. Farmers can make decisions about which species to plant, what to harvest and when and where to sell without waiting for village committees to make such decisions on their behalf. The respective rights and responsibilities of the village collective, on the one side, and households, on the other, were not well defined for many of the responsibility hills. In general, farmers enjoy greater influence over the day-to-day management of forests on responsibility hills compared to the collective management of earlier years. They can determine their own labour inputs to forest management activities, and can influence decisions about species composition, harvesting and sale more easily than before, as the village committee is more sensitive to their pressure and less prone to follow directives from central government.

Nevertheless, decision-making authority is constrained by the continued regulation of the harvest and sale of some forest products, especially timber. In all areas of our study, cutting and transportation permits had to be obtained by farmers wishing to cut and sell wood, no matter what the form of tree tenure. There has been some attempt to ease this administrative burden by allowing township forestry stations to grant permits, as in Libo County. In most counties, however, the county forestry bureau issues such permits, and the journey to obtain them is a burden for villagers, even if officials are disposed to grant the permit, which is not always the case.

## Collective management

Before the 1980s, leaders of a production brigade (now a village) were nominally free to make their own decisions about the management of collective forests; in practice, however, they did little more than implement decisions made from above.

Today, in collective forest management (both modified collective management and shareholding systems), much of that decision-making is undertaken by village committees. In each of the study sites, village committees were active in deciding when and where to establish plantations on collective land and what species to plant there. They also determined when to harvest forest products, subject to obtaining a cutting permit, what proportion to sell and how to divide up the income. In most study sites, village committees were able to enter into contracts that leased land to either villagers or outsiders in exchange for income or profit shares (see Box 2.1). Village committees can also make their own regulations that adapt policy to local conditions. This was the case in Datu village, Libo, in 1990 when regulations were devised to curb tree theft, fire and grazing in newly established plantations.

There remain, however, substantial limits on local decision-making (see Song, 1997, and Sun, 1992, for a discussion of the limitations on local decision-making and forestry). Committees are still constrained by the need to maintain

---

**Box 2.1**  *Contract between the village and government officials*

Four government officials in Libo County leased 26.7 hectares of forest land from Datu Village for a period of 25 years, starting in 1995, to establish a plantation of Chinese fir. They paid 25,000 yuan to obtain use rights to the land and will enjoy all profits from the plantation establishment. They purchased tree seedlings, invited farmers to help plant trees (providing free food for farmers, but without paying for their work), and then hired a household to tend and guard the plantation. The contract agreement stipulated that after the timber was harvested (when the contract expired), the land and some Chinese fir would be returned to the village.

---

good relations with officials at township and county levels. Village committees also are generally unable to resist the larger government-sponsored campaigns that affect forest management. The country-wide ban on logging natural forests, for example, prevented rural communities in Libo from cutting and selling timber. The policy that collectives have rights to take back lands contracted to households if they don't plant trees on such lands within three years made traditional swidden cultivation impossible in study sites in Yunnan.

Devolution may also stop at the level of the village committee, rather than proceeding to the level of household. In the study villages in Gengma, villagers knew little about plans to lease collective forest land to outside developers, or about how lease money was used. In Tiantang Village in Suining, the village committee generally discussed important issues of collective forest management – such as timber harvest and income distribution – with leaders of *cunminzu* and farmer representatives, rather than in larger farmer meetings. Each *cunminzu* selects two to three farmer representatives; usually these representatives are men, and women are much under-represented. The mechanisms for downward accountability are still weak at the village level, and villagers' influence over village committees depends largely upon how well organized and energetic they are (see O'Brien and Li, 2000; Pastor and Tan, 2000; and Oi and Rozelle, 2000, for a general discussion of village-level democracy in China).

## *Self-initiated shareholding systems*

As this is a rather broad category of forest management, the space for decision-making varies substantially. In general, there are two trends.

In the case of cooperation between outside institutions and individuals, on the one side, and local households and village committees, on the other, the outside parties tend to hold most decision-making authority. This was so whether the outside party was a public-sector institution or a farmer from a neighbouring community. Local farmers have a say about income distribution; but nearly all other aspects of management are decided by the outside party, including what to plant, when to harvest and when to sell. The cooperation in plantation establishment between the state-owned County Timber Company and Fujiawan village in Suining County, between the County Reforestation Company and local communities in Jinping County, and between Datu village and Mr Pang, a farmer from a neighbouring village in Libo County, are all examples in this regard. One of the important reasons for the dominance of outsiders is that what the outside party provides is exactly what local households and villages lack in the extreme: capital and technology.

The situation is quite different in the case of cooperation among farmer households. Participating households share decision-making authority, usually on an equal basis, about all aspects of forest management, from species selection, tending and guarding to income distribution. This kind of cooperation is found mainly in Jinping and Libo, Guizhou, but is also seen in Suining, Hunan.

## Devolution and benefits from the forest

To examine whether devolution policies have increased the benefits to farmers from forests, we looked at changes in income and income security, changes in the share of timber revenues that accrued directly to farmers as opposed to government budgets, and the level of asset building associated with devolution.

### Incomes

#### Household-based management

Farmer income and benefits from the forest vary across management types, as well. Researchers were able to collect rough data on household forest income in Suining (for the household management of bamboo) and Jinping (family plots, including former responsibility hills), but found the issue quite sensitive in other areas. In Suining, incomes were reported to have risen dramatically after household management of bamboo was introduced. According to farmers interviewed, their income from bamboo shoots is now double that in the past. Other factors influenced income in this case, particularly the citing of a bamboo-processing plant in the county; but tenure reform was certainly a contributing factor. In Jinping, too, farmers reported substantial increases in income from trees of Chinese fir that were distributed to them or that they planted on family plots.

These changes in income occurred for at least two reasons related to devolution. Firstly, farmers felt that their tenure rights over trees in family plots and over bamboo forests contracted to them for 30 years was more secure (see Sun, 1992; Bruce et al, 1995; and Zhang, 2000, for a general discussion of forest tenure security in China). They had control over the terms of the contract, which affected their sense of control over the resulting income. Villagers saw this as a benefit in itself. It also spurred people to plant species that they would not have otherwise planted, which led to increased income generation. Farmers in Libo, for example, reported a much greater willingness to plant Chinese fir after the reforms, citing greater tenure security as a major factor in their planting decisions. Similarly, in Chuxiong, farmers planted eucalyptus in much larger numbers after the reforms, comfortable that they could realize the benefits from sales of oils and timber.

Secondly, family plots (including former responsibility hills) and bamboo contracts also allocated a greater share of lucrative timber revenues to farmers (see Table 2.2). This proved a strong incentive to farmers to plant and maintain timber species, as reported by farmers in Jinping. The combination of greater tenure security and greater shares of timber revenues linked farmers' incomes more closely with their management performance, and helped to achieve some of the gains in forest cover experience in southern China during the 1990s (see Albers et al, 1998, and Yin, 1998, for alternative explanations for forest cover increases).

Under responsibility hills, there is not only a sharing of responsibility for management, but also of income between the collective and individual households. The amount of income earned by each household varied and was

**Table 2.2** *Distribution of commercial timber revenues by management category among local interest groups, Dicha Village, Jinping*

| Tenure category and income recipient | Percentage of timber income received by different members of community (%) |
|---|:---:|
| Household management: | |
| • village collective | 10 |
| • village households | 90 |
| | |
| Shareholding systems Ximen Forest Farm: | |
| • village committee | 10 |
| • forest farm | 3 |
| • village households | 87 |
| Gaogang Forest Farm: | |
| • village collective | 10 |
| • forest farm | 20 |
| • village households | 70 |
| | |
| Self-initiated shareholding systems (farmer–farmer collaboration): | |
| • village committee | 10 |
| • parties establishing and managing plantations | 40–50 |
| • households | 50–40 |

*Note*: timber income in China is distributed between outside interest groups (government administration and the forestry department) and local villages and households. Generally, the local interest groups obtained about one third of a timber sale's value as its income (see Table 2.5). This proportion of income is counted as 100 per cent in this table.
*Source*: compiled from data collected by the Jinping team

determined based on terms of 'contracts' between village collective and households.

## Collective management

The benefits of collective management were primarily from timber revenues, which were distributed among local development projects, the village committee, local schools and individual households in the village. The support for village-level institutions can be thought of as indirect income to households because they no longer have to bear these expenses (or not all of them) without giving up village services and development projects. The remaining income is distributed directly to households (see Tables 2.2 and 2.3), with each villager receiving an equal share.

Individual households benefited mostly indirectly from improved management, compared with household management systems. Hence, the level of village income under collective management varied primarily with the quality of management by the village committee. In Tiantang village, Suining, the forest was managed well, and in recent years the village derived great income from timber (1.35 million reminbe (RMB) in 1999).

Table 2.3 *Distribution of benefits from commercial timber within the village under modified collective management: Tiantang Village, Suining*

| Tenure category and income recipient | Percentage of timber income received by different members of community (%) |
| --- | --- |
| Modified collective management: | |
| • expenses on reforestation and village infrastructure, and salaries of village committee and teachers | 30 |
| • village households | 70 |

*Source*: compiled from data collected by the Jinping team

## Self-initiated shareholding systems

Under these management institutions, the parties work out themselves a scheme for distributing benefits. The scheme is specific – parties involved are clear about the proportion of income they can get from their joint tree-planting efforts, and terms generally must be acceptable to all of them. The company–village partnership in Fujiawan village, for example, divided income from timber according to the following terms:

- first crop: company 60 per cent and village 40 per cent;
- second crop: company 55 per cent and village 45 per cent;
- third crop: company and village 50 per cent each (see also Table 2.2)

Under these schemes, the sense of security of income is often high because of villagers' direct involvement in, and knowledge of, the contracts. This confidence can be misplaced if there is little chance that outsiders can be brought to justice for breaking contracts, as was the case in some of our study sites in Gengma. With farmers increasingly demanding fair protection from Chinese courts, however, the outlook for these shareholding systems is likely to improve.

## Participation in management of state forests

Villagers or communities participating in the management of state forests did not enjoy the full scope of decision-making rights that they did in household-based management and collective management. They did benefit from increases in income. If villagers take responsibility for forest guarding and management, the forestry department allows them to collect non-wood forest products, tap resin, gather fuelwood from thinning branches and harvest deceased and decayed trees. With approval of the forestry department, villagers have even been able to harvest small amounts of timber for on-farm use. These use and disposal rights save time and money that would have been spent on finding other sources of forest products, and sometimes generate income through direct sales of forest products collected. The income benefit is not large, but farmers in Jinggu were generally quite pleased with the change in their access to state forests.

## Village and household assets

Devolution policies have also allowed villages and farmers to use trees and forests as assets, increasing the potential long-term gains in income and structural changes in livelihoods. Assets provide for security against future needs and flexibility in responding to market opportunities, and can be used as collateral to secure loans. In our study sites, village committees and individual households have both benefited from using trees in these ways.

In general, farmers benefit directly from trees as assets when the trees are located in family plots, and indirectly when the trees are part of collectively managed forests. On average, for example, each household in Datu Village, Libo, has about 10 hectares of Chinese fir established by the household itself on so-called wastelands. Part of this forest land was distributed to households for use as family plots during the early 1980s, though most was distributed when plantations were planned during the late 1980s. These plantations can be quite valuable. One farmer has 20 hectares of Chinese fir that is valued by the forestry department at about 500,000 yuan (approximately US$60,000) after eight years of growth. Two other farmers, jointly managing their lands, have 23 hectares worth 743,000 yuan (US$90,000). Such assets can be converted into income in times of need or when market prices are favourable since farmers have been allowed to sell live stands, and do not have to wait to obtain a cutting permit to sell. Perhaps the most important use of the asset today, however, is as collateral to secure loans. Farmers in Datu and Shuiwei have used plantations as collateral for loans from commercial banks. These have financed animal husbandry and trucking businesses for farmers, as well as school construction. Bamboo has been used as a household asset in Suining, as well.

Of course, devolution policy is not the only factor to contribute to increases in the assets of villages and households. There are some other factors, such as improved extension and credit services and market information provided by public sector. Devolution policy is an important factor according to farmers in our study sites, however, and asset-building must figure in an evaluation of devolution policies in China.

## Increases in forest resources

Devolution policy contributed to an increase in forest resources in terms of area and total stock in the long run. There was a dramatic decline in forest stock in many places immediately after the introduction of household-based management during the early 1980s. Many foresters and academics believe that it was the second worst deforestation in China over the last half century (the first worst deforestation took place during the Great Leap Forward in 1958 to 1960). Yet, the trend seems to have reversed itself afterwards, and forest resources are increasing steadily as farmers have become confident about tenure security and the benefits that they can derive from forest management. At national and provincial levels, and in all our study sites, forest area and stock have increased since the 1980s, especially since the late 1980s (MoF, 1989, pp121–123; MoF, 1994, pp93–96). For example, the stock of non-state forests in the 12 provinces noted earlier increased from 2456 million cubic metres during the 1984 to 1988 inventory period to 2585 million cubic metres during the 1989 to 1993 inventory period (MoF, 1990, pp106–132; MoF, 1995, pp85–105). Trends in the research sites, shown in Table 2.4, are similar. Increases in non-timber plantations are even more encouraging. The increases resulted from a combination of many factors. However, it is safe to say that devolution policies are one important factor that contributed to the increases because much of the recovery is the result of an expansion in plantations, which are closely linked to the improvement in farmer incentives associated with devolution policies.

The impact of devolution policy on forest quality in terms of stock per hectare, as well as on biodiversity, is controversial. National inventory data show that stock per hectare declined in China over the last two decades. Some link the decline to devolution policies; but both national forest inventory and surveys at our study sites do not show a causality of the decline and devolution policies. The general trend over the period was that mature forests were (over) harvested and new plantations were established. As a consequence, stock per hectare declined.

There are also claims that devolution policies have caused biodiversity loss. Mixed broadleaf forest declined in volume by 54 per cent during the 1980 to 1988 period, while 'cash' forests of monocrops increased during the period by

**Table 2.4** *Changes in forest resources*

| Village | Forest area: hectare (per year) | Forest area: hectare (per year) | Percentage change (%) |
|---|---|---|---|
| Dicha, Jinping County, Guizhou | 504 (1983) | 1621 (1995) | +222* |
| Datu and Shuiwei, Libo County, Guizhou | 152 (1984) | 1064 (1994) | +600** |
| Suining County, Hunan | 164418 (1983) | 179707 (1990) | +9* |

*Notes:* * Mainly achieved through tree planting.
         ** Plantation forest only
*Source:* compiled from data collected on field surveys

more than 21 per cent (Albers et al, 1998). Yin (1998) predicts a continued loss of larch, Korea pine, ash, walnut and other primary species as they are replaced with plantation species such as Chinese fir and poplars. Communities and households with increased decision-making authority may prefer monocultures of valuable trees in establishing plantations. Examples from our study sites include the Chinese fir plantations in Jinping and Libo and the eucalyptus plantations in Chuxiong. However, this preference is not the cause of biodiversity loss in our sites because biodiversity had already declined before monocultural plantations were established, largely on wastelands or degraded collective plantations. Moreover, in some places where natural forest has persisted, the new policies have not affected existing practices that protected key environmental services. To the extent these services are linked to biodiversity, local people have been able to also protect biodiversity.

This has been the case in the Ailao Mountains of south-eastern Yunnan, where Dev Nathan conducted fieldwork, Datu Village, Libo; in Dicha Village, Jinping; and in Chuxiong, where farmers placed a high value on the environmental services of forests, particularly those related to watershed protection. Changes in forest tenure there have not led to a significant degradation of broadleaf forests. Furthermore, in cases such as Jinping, farmers have agreed to leave part or all of the natural broadleaf forests under the control of the collective to ensure the provision of ecological services.

## Management costs

While villagers benefited from policies supporting forest management by households, in many cases, management costs increased. In an attempt to be fair to all households in a village, collective forests and forest lands were divided into many tiny plots of different quality before being distributed to villagers. Plots were divided according to species composition, tree age and density, soil quality, slope, and distance from the village, among other criteria. These plots were then matched to families of different sizes. A family often received five or more scattered forest plots of different kinds of forest that together were roughly equivalent in quantity and quality to the forest holdings of all other villagers. Any given forest patch thus had many managers. As a result, forest boundary-marking, tending, guarding and logging were difficult and costly to carry out (Sun, 1992; Liu, D, 2001).

In study villages in Chuxiong, for example, villagers said that it was difficult to guard their family plots and responsibility hills under devolved management. Under collective forest management, a village needed only two or three forest guards for its entire forest area. After household-based management was introduced, each household had to allocate a person to look after its forests, though not necessarily on a full-time basis. Households with fewer labourers found it especially difficult to bear the labour burden. In response, households attempted to coordinate their efforts. They guarded their forests by shifts: one household guarded forests one week and then transferred the responsibility to the next household. Nevertheless, this was more costly than guarding under collective management had been. The failure to keep up with forest guarding also increased opportunities for tree theft.

The high costs and operational difficulties associated with household-based management help to explain the development of both self-initiated shareholding systems and land leases. These systems allowed farmers to compensate forest guards with larger shares of forest revenues and assets, or to contract others to protect the forest.

## Growing inequality among farmers and between localities

Household management of forests has contributed, in combination with other factors such as the liberalization of the farming sector and differences in human capital, to increased inequalities among households. Gustafsson and Wei (2000) note rapid increases in income inequality at the national level, and within the areas officially designated as poor in the west. Our own study found evidence of rising inequality within and between villages that appears to be directly linked to forest policy reforms. Within villages in Libo, for example, households that had effectively equal access to forest lands through their place in the collective now might own as many as 30 hectares or more of forest, or as little as 2 or 3 hectares (also see Table 2.5). As yet, the differences among households are small relative to those in many other countries. Furthermore, the benefits of household management for even poorer households appear large enough to maintain broad support for the policy among villagers. Still, within the Chinese context, such differentiation is cause for concern among villagers and government officials alike.

In fact, there have been complaints, particularly where collective forest lands have been leased to individuals or companies. In two villages in Suining, for example, farmers became angry enough that they pressured a neighbour who had leased nearly 200 hectares from the village to return most of the land.

**Table 2.5** *Wealth of farmer households and their plantation establishment in Datu and Shuiwei, Libo*

| Wealth rank | Average annual income (yuan) | Plantation establishment (hectare per household) | |
| --- | --- | --- | --- |
| | | Datu Village | Shuiwei Village |
| Wealthy | Over 10,000 | 10 | 6 |
| Middle | About 5000 | 7 | 3 |
| Poor | About 2000 | 4 | 2 |

*Source:* compiled from field data

In study villages in Gengma, farmers complained when collective forest land was leased to a public institution for the development of cash crops. When the cash crops failed, villagers received nothing, not even wages for labour. The terms of the lease were never clear to villagers, and they had no option for claiming compensation. Land in Gengma was also leased to a retired government official for orchard development. Again, villagers were unclear about the terms of the lease, and had given up access to collective forest land without any assurance of how they would be compensated.

Inequities can also arise within the household. In Datu Village, Libo, decisions on species choice, harvesting and marketing are sometimes made after family discussions, but often by household heads acting on their own. Decisions on joint plantation establishments or collective forest management are usually made by a representative of the household, often the household head. Researchers did not ask much about gender inequalities within the household, or how devolving authority to households or villages affected women's benefits from forests. This issue requires further research. Initial findings from the fieldwork conducted by Dev Nathan and Govind Kelkar in Yunnan, however, indicate that women's preferences for the species planted in family plots, and the use of forest income, were often ignored, with negative implications for women's well-being.

## *Building the capacity of communities and households to manage forests*

Villages and households had to develop new capacities to manage forests once they moved away from the collective system. There is strong evidence from our 15 study villages to indicate that they have done so. Devolution policies have contributed to building capacity both by creating a need for it – particularly by introducing household management systems – and by facilitating opportunities for villagers to learn.

Villages and households have developed their capacity to manage forests, in part, by improving their links with outside organizations and individuals who can provide them with the technical and market information that they need to realize greater benefits from their forests. This was evident in the treatment of our research teams. Households interviewed in Datu and Shuiwei, Libo, in particular, showed great hospitality to team members, and were keen to exchange information about forest management opportunities with us. Ms Yang, party secretary in Tiantang village, Suining, made a typical request that the research team provide her and other villagers with more forest technical advice and with market information whenever team members visited there. Beyond our own research team, villagers reported taking greater initiative to seek out government officials, companies and foreign investors for forest management support. Village committee members in our sites in Gengma, for example, were actively seeking advice from government officials, media sources and traders on which forest products might be profitably sold through the special trade arrangements along the Myanmar border.

Devolution policies have made it easier for villagers to learn from outsiders. For example, government foresters and forest researchers are accessible to individual villagers under household management systems – not just their representatives in the collective – and the opportunities for villagers to learn from these 'outsiders' are increased. Policies that have encouraged the lease and transfer of live stands have also allowed farmers to partner with outsiders with more technical knowledge. While villagers primarily provide labour under many of these lease and transfer agreements, in other agreements, villagers gain experience in technical and marketing aspects of forest management.

Devolution policies have also helped to create an environment in which villagers enjoy greater freedom to experiment with forest management arrangements on their own. The most obvious example is in household management schemes, where farmers are trying out new species, especially fruit trees and other species that are quick to mature. Such experimentation was found in each of the study sites where household management was found. Other experiments focus on developing new institutional arrangements (see Yeh, 2000, for an example). Self-initiated shareholding systems are the result of institutional experimentation and learning by individual farmers. Many households interviewed in Jinping and Libo were involved in such systems, both among themselves and with outsiders. As suggested above, this kind of shareholding addresses critical problems related to the fragmentation of forest plots and to the lack of capital in many villages. While formal shareholding policies have not allowed for much experimentation on the part of villagers, they have provided a model for developing the informal shareholding systems. The tenure rights enjoyed by individual farmers during the preceding decade were a prerequisite for farmers' interest and confidence in taking on such an experiment. As a result, although self-initiated shareholding systems are not the result of formal policy, changes in policy helped to create the conditions that allowed them to develop.

Villagers in many locations also learned to protect their interests using laws and policies, developing new kinds of political capacity. Farmers in Shuiwei Hamlet, for example, have developed new strategies to resolve conflict between them and the Shuiwei Village Committee over land ownership. During the commune era, a Chinese fir plantation of over 3 hectares was established by the village committee, and the committee sold trees in 1993. After the trees were sold, the hamlet wanted to take back the land and plant trees again. They claimed that they had the use rights to the land because the land was distributed to them during the early 1960s. However, the village committee also claimed the land. The conflict was not resolved through negotiation and conciliation, as would normally have been the case. Instead, the case will go to court. In rural China, especially in isolated areas, it has been nearly impossible for villagers to go to court with their village committee. In other villages, farmers were using the media to criticize local officials and others who threatened their interests. While devolution policies are not solely responsible for this development (Oi and Rozelle, 2000), farmers' sense that they have secure tenure rights has certainly been a contributing factor, as have farmers' contacts with outsiders and experiments with new institutional forms.

## FACTORS THAT AFFECT DEVOLUTION IMPACTS

Identifying the impacts of devolution policies does not, in itself, explain why they worked, or provide much guidance for future policy reforms. We therefore look at some possible explanations for the impacts discussed above. Of course, any policy impact is a result of many factors that work together, and it is often difficult to take account of all of them or of their relationship with one another. We provide several explanations in the following sections. They are presented in the order of their importance in affecting devolution impacts, based on our assessment.

### The transfer of comprehensive rights

Responsibility for the management of collective forests in China has been transferred from the forestry department to village committees, and from collectives to households. There has not, however, been a full transfer of tenure rights in some cases, particularly rights to harvest and market timber. In all study sites, the timber harvest has been highly regulated using cutting quotas, cutting permits, transport permits and processing permits, which is consistent with observations from other parts of China (MoF, 1990, pp37–39; MoF, 1987, pp477–478; MoF, 1988, pp11–12, 485–486; MoF, 1996, pp26–29; MoF, 1990, pp40–41, 46–47). Such partial devolution creates conflicting incentives and limits local enthusiasm for tree planting and sustainable forest management. Farmers were more interested in bamboo in Suining, for example, because bamboo harvesting was much less heavily regulated and taxed than timber (see discussion under 'The consistency of devolution policies and other forest policies', and Chapter 5 'Taxes and regulations'). Villagers were not keen to plant Chinese fir (a timber species) in Libo before the trading of living trees and plantations was allowed because obtaining harvesting permits was such a burden. For the same reason, households in Chuxiong were most enthusiastic about eucalyptus trees, from which they harvest leaves for oil extraction, and about fruit and nut trees. Villagers who were interviewed indicated that they were not active in planting timber species because they were unable to harvest and market timber when the need for income arose. Instead, they depended upon calendars and cutting plans established by the government. Forest use rights are important; but rights to harvest and dispose of forest products are equally important in encouraging tree planting and protection.

## The link between farmer benefits and farmer efforts in forest management

The Chinese government has increasingly adopted a position that granting greater tenure rights to forests leads to greater wealth creation among farmers (Grinspoon, 2001), primarily by improving incentives for investments in forests. We found, in general, that when there was a closer link between farmers' investments in forestry and the income that they could earn from those investments, forest management improved. However, the link was stronger or weaker depending upon the type of tenure reform considered, as well as other factors, such as taxes, regulations and markets (Daowei, 2001). We examine farmers' incentives in the various forms of forest tenure below.

### Household-based management

The shift in management rights and responsibilities, especially to the house-hold level, has increased the linkage between people's investments in forest management and the benefits that they receive from forests, leading to increases in the quality and extent of forests during much of the 1990s, as discussed above. Among the devolution policies in place, household-based management has provided farmers with the greatest share of timber income and has enabled them to build the largest assets from forests. In four out of six counties studied, household management systems experienced the highest levels of investment in reforestation and other forms of forest management. In the study villages in Libo, for example, all households planted Chinese fir on wastelands distributed to them. Similar stories can be found in Jinggu and Jinping, as well as for bamboo management in Suining. The investments were sometimes substantial. In one of the study villages in Chuxiong, a farmer leased land from the village collective during the 1990s and arranged for his two sons to work full time, looking after the existing pine forest and planting nut and fruit trees on unforested land.

### Collective management

Collective management conferred benefits that were less directly linked to the efforts of individual households. Although, in some cases, revenues from collective management could be distributed to individual households, most often, households did not receive significant income or assets from collective forests. Frequently, revenues were used to pay for services that were, in principle, valued by farmers. The poor accountability of collectives to farmers, however, meant that the money was often used for projects and programmes that were of little interest to farmers. In almost all of our study sites, collective forests have depended upon investments made by the village committee, rather than by farmers themselves. Not surprisingly, then, many of these forests have done poorly, as committees do not have the capital or human resources to manage forests on their own.

This has tempted some committees to contract collective forest management to outsiders. Occasionally, this has had disastrous results for the forests and local farmers, as in Gengma, when outsiders break contracts or committees permit forest uses that do not take into account forest use by locals, especially the very poor (see Lu et al, 2001, for a general discussion of some of the shortcomings of these contracts). We note that devolution policies have created an array of new incentives for both individual households and village committees that can leave the poorest forest users – those who still depend heavily upon rights to collect fuelwood and other forest products from collective forests – without access to the resources that they need. Better-off farmers can pursue the opportunities associated with household management or tree leasing without fear of what happens to the collective forest. The village committee can dispose of the collective forest as it desires without substantial opposition from better-off farmers. This leaves the poor with fewer chances to benefit from household management or leasing, and with little influence over the village committee's forest management decisions.

## Self-initiated shareholding systems

Self-initiated shareholding systems represent a broad category of forest tenure and management arrangements and are, as yet, relatively undeveloped. It is difficult to explain what is happening in them with any certainty, but we offer some preliminary suggestions.

Under farmer-to-farmer contracts, profits from the sale of forest products are distributed directly to the contract participants. Farmers receive a share based on their investments in forest management – capital, land and, in some cases, labour. Farmers are encouraged to invest and improve their management performance to earn larger incomes. This form of management is rather common in Jinping (see Table 2.2) and enjoyed substantial farmer support.

Under company–village contracts, income distribution is direct for both the company and village, but indirect for households in the village. Outside parties take responsibility for daily management in most cases, while the village contributes land. The share of income enjoyed by each is based on its investments in forest management and through negotiations over the relative value of these investments. The village committee participates in negotiations on behalf of households, and organizes local capital, land and labour inputs.

Households in the village have little direct say about the contract or how the forest is managed. As in areas of Gengma, village committees that are not accountable to farmers can negotiate contracts that bring few benefits to the local community, and generate little interest among farmers.

## The consistency of devolution policies and other forest policies

The impacts of devolution policies are closely linked to other forest-related policies. If these other policies are not consistent with, or supportive of, devolution policies, devolution is less likely to result in better forests or greater benefits and decision-making authority for local forest users.

Policies regulating the distribution of income from forest product sales and forestry investments are particularly important in this regard (see J Liu, 2001). Data show that outside interest groups (such as the forestry department and government administrations) captured over 50 per cent of timber sale value in the form of taxes and fees, while timber merchants often captured another 15 per cent. This left about one third of the sale value for village committees and/or households to cover the costs of planting, tending and thinning, guarding and harvesting timber in Suining and Jinping (see Table 2.6). This sort of income distribution is common in the other study sites and throughout the country, as well (Wu, 1993; Wang and Yang, 1994). This has encouraged farmers to spend more time in maintaining their bamboo plantations and fruit orchards, where taxes and fees are relatively low, and where farmers capture most of the benefits. This explains, at least in part, why in Fujiawan village, Suining, every household is active in bamboo forest management and bamboo processing, but why Chinese fir plantations had to be established in collaboration with the county timber company.

Forestry department policies regarding reforestation investments also limit the potential positive impacts of devolution. For example, a policy requires a minimum land holding of 20 hectares to access low-interest government loans, free tree seedlings and other forms of government support for reforestation. The requirement does not favour household-based management because few households have land holdings as big as that. Farmer households have to collaborate in establishing plantations if they want to obtain the funds. Indeed, villagers in Jinping confirmed that this is one reason for them to form a shareholding group.

A more dramatic example is found, of course, in the logging bans imposed on natural forests within the Yangtze watershed in 1998 (Shen, 2001). In response to flooding during the late 1990s, the central government imposed bans on logging in natural forest areas of the watershed, including some areas where household management and other forms of devolution had taken root. There is fear among many foresters that the ban will discourage farmers from investing in forest management on their household plots, and hasten the shift to plantations and non-timber tree species.

**Table 2.6** *Distribution of benefits from commercial timber between community and outside stakeholders*

| Costs, tax and fees, and community's share of income | Percentage of timber sales value captured (%) | |
| --- | --- | --- |
| | Dicha Village, Jinping | Suining (1997–1999) |
| Income share of stakeholders outside of community: | 51 | 51 |
| • tax and fees | 50 | |
| • fees for township | 1 | |
| Costs on logging, transport and merchant's profits: | 16 | 15 |
| • logging and transport | 10 | 5 |
| • merchant's profits | 6 | 10 |
| Income share of village* | 33 | 34 |
| Total | 100 | 100 |

*Note:*\* The village's share of income is not the same as its profits or net income. This income must be used to cover costs on planting, tending and guarding. Some studies estimated that villagers have no net income from timber sales. Instead, they receive only a small salary for their labour for planting, tending and guarding.
*Source*: compiled from data collected by the Jinping team

Other policies, such as those improving the accountability of village committees to villagers, supporting contracts in the legal system, and improving the infrastructure for forestry have helped to make devolution more successful in some areas. In Suining, for example, the decision to locate a bamboo processing plant in the county provided farmers with a ready outlet for the bamboo over which they now had control. Income rose dramatically as a result, and farmers were quite content with their new tenure arrangements.

## Village capacity for cooperation

Villagers' capacity for cooperation among themselves or with other groups affected their ability to wield influence over local officials, manage internal differences of opinion and create shareholding agreements, and thus to take full advantage of devolution policies. We look here at the different types of social capital that influenced devolution outcomes. In this chapter, we define social capital as a set of social resources possessed by households and communities, the presence of which enables them to work together effectively towards the attainment of certain ends, as defined by Loury (1977), Coleman (1990) and Bourdieu (1986). Specifically, social capital refers to the norms, social networks, alliances, voluntary associations, mutual obligations and information potential that facilitate coordination and cooperation among individuals and groups (Putnam, 1993).

We look here at four aspects of cooperation that influenced devolution outcomes: cooperation between private and government actors; alliances among households; trust among members of a community; and norms that influenced how national minority communities valued forests, leadership and abided by rules.

## Public–private cooperation

Many employees of grassroots-level government are from rural communities. They still have strong ties to their relatives and friends in the countryside and communities from where they originally came. They have a good knowledge of villagers' desires and interests, and of the social and environmental situation in their communities. On the other hand, they work with government agencies and have access to government resources and information; some of them are policy-makers. Such government employees are a good channel through which to feed community views and opinions, and they are often able to help create or expand space in decision-making regarding forest management for local people, especially when they have common interests with rural households (see Evans, 1996, and Joshi, 1999, for a general discussion).

Local people can take advantage of this to become better informed about policy, ask for exceptions or seize marketing or investment opportunities quickly. Good working relationships were established between villagers and county officials in Suining, leading to improved marketing opportunities for bamboo. Local officials were also strong advocates for the interests of farmers in southern Yunnan, helping them to gain access to state forests for limited but important forest products. Unfortunately, not all villagers are involved in such personal networks, and social capital has often worked to the disadvantage of some sections of the population, even while benefiting others (see Box 2.2). There is a need to develop and consolidate more inclusive alliances with government officials to ensure that more people benefit from devolution policies. A lack of such capital already helps to explain some of the differences in policy impacts among villages and among individuals within single villages.

## Alliances among households

Alliances among farmer households are another expression of social capital that influences devolution policy impacts. Alliances can be temporary or relatively permanent, formal or informal, and can affect any number of management activities, from investments to forest protection to resolving disputes among local farmers.

Many residents in Fujiawan village, Suining, for example, formed a temporary alliance to fight for their interests in the distribution bamboo forests in 1997, against the interests of village leaders and the stated letter of household management policies. Residents who had received small allocations of private household plots during the 1980s, from 0.2 to 0.3 hectares, demanded that they

---

**Box 2.2** *Emerging inequalities*

Mr Pang, a Miao man in Shuiwei Village, Libo, is 28 years' old, has a middle-school education, is able to communicate well in Mandarin and has a good relationship with government officials at township and county levels. With this background, he is able to obtain good information about government policies and market conditions. He was one of the first in his village to establish a Chinese fir plantation, and now has 20 hectares.

In sharp contrast, Mr Y, a Shui minority in the same village, is 45 years' old. He is illiterate and poor. He has limited knowledge of policy or markets, is unsure if the new devolution policies will remain in place, and doubts the security of his tenure over forest lands. As a result, he has planted only 2.3 hectares of Chinese fir. He now wants to develop more; but there are no more wastelands available at this time.

---

be compensated with larger shares in the distribution of bamboo forests. Village leaders, who had received 2 to 3 hectares in the earlier distributions, argued that such redress was against the central government policy. Based on common interests and clan linkages, those households with smaller private plots prepared a document, signed by the heads of all households involved, that described the situation from their point of view. They then appointed a capable villager to send the document to township government and county government, respectively. Their demand was supported and approved by the county forestry bureau and the township government in just a few months.

Similarly, in Lailaishui village, Suining, one farmer acquired more than 200 hectares of forest land from the village committee. His contract granted him use rights to the land for 30 years. His neighbours were very angry as there was little wasteland left over for others to obtain on contract. They pressured the farmer until he decided to return all but 12 or 13 hectares to the village, with the trees that he planted included. His wife commented: 'Our life in the village will be very difficult and even a disaster if we don't return land and trees to the village, and all other villagers will make enormous trouble for us.' These examples suggest that policy impacts will be better if villagers have sufficient social capital to oppose abuses and propose reforms in how policy is implemented.

## Trust

In contrast, villagers from national minority villages in Gengma had not been able to cooperate effectively in opposing forest management decisions against their interests. We were told of individual efforts to make complaints about the unfair contracting of collective forests; but there had been no large-scale collaboration to address the problem. The relative poverty and isolation of these villagers may help to explain their inability to cooperate; but a lack of social capital must also have been at work. People in Datu and Shuiwei villages, Libo, for example, live harmoniously; but cooperation among them is limited to simple forms of labour exchange. There has been cooperation among several households; but villagers are not willing to cooperate at a larger scale – for example, village level or above. In our discussions with them, it became clear that their reluctance was due to past experiences with large-scale cooperation, often forced upon them by government policy. They illustrated the point with a story of a plantation established through villagers' labour during the commune era. The plantation was later taken over by the township government, and villagers received nothing for their efforts. Similarly, a collective plantation developed during the commune era in Shuiwei Village was later sold by the village committee. The money was then invested in coal-mining enterprises and was lost when the mining failed. Villagers again received nothing. This has made it difficult for village-level cooperation to take root again in Datu and Shuiwei. Other villages have had similar experiences and are equally shy of village-level cooperation.

## National minority cultural norms

The introduction of household-based forest management did not induce significant deforestation in some areas, largely because of unique local social capital: national minorities' cultural norms and the positive role of their leader. In the case of Datu Village, Libo, local Miao cultural norms encourage honesty and place a high value on forests. Miao people live harmoniously and do not take anything belonging to somebody else – there is no stealing and no need to lock doors. They do not cut trees from another person's forests without agreement or permission. These norms mean that little or no forest guarding is necessary, all but eliminating one of the important costs of household management systems, and one of the principle barriers to the success of that form of devolution. Moreover, forests play an important role in Miao culture, and management traditions have developed over a long history of relations with the forest. Each Miao village has a holy forest around it that is well protected (see Box 2.3). More importantly, the Miao also apply the principles that protect holy forests to other forests – limiting harvesting, encouraging regeneration and restricting access.

Minority leaders can play a critical role in maintaining useful social capital within Miao and other communities by resolving community disputes. The Miao leader in Datu Village, Libo, is an exciting example (see Box 2.4). His influence is felt at village level and beyond, and allows the cooperation among the Miao to take place at scales above the village, even crossing county boundaries. Disputes are not restricted to forest issues; but forest disputes are definitely among those that can be solved by ethnic leaders. In this sense, the role of the Miao leader (or other similar ethnic leaders) is a form of local governance built on strong, ethnically oriented social capital.

---

**Box 2.3** *Miao holy forests and forest management*

It is said that Miao ancestors fought with another tribe in central China in ancient times. At the beginning of the conflict, they were winning. Later, their enemy tricked the Miao leader, and Miao people lost the war. The leader died in a forest and his people did not have time to bury him because their enemy was approaching. They simply covered his body using tree leaves. Thirteen years later, when peace returned, Miao people found his body in the forest and buried him. They were grateful to the forest for covering the leader's body for such long a time. They treated the forest as a holy forest and held sacrifices and other rites for both the leader and the forest. Since then each Miao village has set aside a holy forest, without exception, and the rites have been held in that forest every 13 years. Domestic animals are not allowed to enter the holy forest and trees are not to be cut. This reverence for the forest explains, in part, why significant deforestation did not take place in Datu and Shuiwei villages after the introduction of household-based forest management.

---

**Box 2.4** *The role of the Miao leader*

The Miao leader is informally elected from villagers based upon his reputation and is not appointed by government. He is thought to be just and unselfish, and he generally enjoys the respect of the Miao people. Consequently, his decisions are often more acceptable to Miao people than those of government officials. One of his primary roles is to deal with conflicts among villagers and between communities, including land disputes, forest access and use conflicts, as well as marriage problems. For example, a hamlet from a neighbouring county planted Chinese fir in the boundary area in Datu Village in 1994. Government civil affairs bureaus of both counties tried several times to solve the ensuing conflict, but failed to win approval for their proposals from farmers in either village. After some consultation, both sides agreed to request that the Miao leader in Datu Village mediate the discussions. His involvement helped to reach an agreement that was acceptable to both sides.

---

## The role of local government

The transfer of responsibility and rights for forest management does not necessarily mean government retreat. Until civil society capacities develop further, government, especially local government, can be very influential in the lives of forest users, with both good and bad outcomes. Local government can 'localize' national and provincial policies, adapting the specific details to local management practices, environmental conditions and social relations. This is quite often the case in areas where national minorities with little influence on national policy become majorities in local administrations. Local government also plays a role in resolving forest boundary conflicts between communities, establishing factories or enterprises that use forest resources as raw materials and that add their value, providing quality technical and financial services and market information services. In contrast, local governments can sometimes compete with villagers for forest revenues and may be involved in unethical contracts with outsiders. Much depends upon whether they can be held accountable for their actions by local farmers, though fiscal incentives and social networks can also play an important role.

Localization can be a powerful influence on the lives of forest users, and the starting point for more local control of forest management. In Jinggu and Gengma, local government contracted/commissioned state-owned forests to local forest users to allow local use of state forests. Farmers were compensated with a very small sum of money for forest management; but they were given rights to harvest resin from forests and sell it directly to a state-owned processing plant, as well as to harvest other non-timber forest products and fuelwood. Similarly, in Yi villages of Lashi Township in Lijiang County, township officials allowed Yi villagers to cut and sell timber even though the township had no quota for logging, and thus no legal right to do either. Officials justified the decision by pointing out that the Yi had little cash income and that much of what they did have was related to timber production. The Yi needed the money for school fees and other cash expenses.

Forest management in Suining County provided an excellent example of localization of central policy. The government facilitated the recollectivization of forests in 1987 that were once under the management of individual households (as responsibility hills). This was in response to the concern of officials and farmers about increased deforestation during that time. Ten years later, the government redistributed collective bamboo forests to households through contracts with a term of 30 years, based on its belief that the link between benefits and local management would be greater with bamboo.

Local government can also be very influential in the lives of villagers by facilitating the development of factories that use forest resources as materials. In Suining, county government helped to establish several factories to make bamboo floor tiles, and facilitated the distribution of first-stage bamboo processing equipment. As a result, local farmers earn much better incomes from bamboo forests than from simply selling bamboo poles, and have a ready market for their production. Local government officials are now looking for better markets for bamboo shoots as a secondary source of income, and for ways of controlling the increases in erosion that have followed more intensive management of bamboo stands. In Jinggu, the forestry bureau established a large resin-processing factory, which uses resin that farmers tap from their family plots and from state forests that they manage. Resin tapping, rather than timber, is a major source of cash income for many households in the county.

Besides better use of forest resources, local government also played a critical role in helping villagers to develop the technical and managerial capacities required to establish plantations. In Datu and Shuiwei villages, Libo, devolution of forest management from village to households improved the management of existing forests; but large-scale plantation development did not come until the government forestry department was involved (see Box 2.5; see also Albers et al, 1998). The example argues for a strong government role even after devolution.

Even though the county forestry bureau offered some training, the skills of villagers in tending and thinning Chinese fir were still limited, which reduced the potential of Chinese fir as a source of income. Furthermore, local villagers lacked access to market information. As a result, they often sold their stands or trees at prices much lower than market prices. The experiences in Datu and Shuiwei clearly indicate that government can and should continue to support farmers in the areas of planning, providing capital and offering technical and market advice, even after devolution policies are implemented.

---

**Box 2.5** *Plantation development in Shuiwei Village, Libo*

A total of 2340 hectares of Chinese fir plantations were established in Shuiwei Village and surrounding areas during 1989–1993, well after forest devolution policies were first introduced in the area. The plantation development was part of a poverty alleviation project of the former Ministry of Forestry of China (now the Chinese State Forestry Administration), with the ministry providing an interest-subsidized loan totaling 910,000 yuan. The Libo County Bureau of Forestry was responsible for loan repayment and for organizing project planning. The Bureau distributed funds to farmer households in the form of both cash and seedlings. Farmers planted trees separately on their own lands. The trees are the property of individual households. The farmers will have to repay loans to the Bureau at 7.5 to 15 cubic metres per hectare in 15 years.

The Bureau developed a general plan for the plantation, and relevant township governments organized the implementation of the plan. Township forestry stations (the lowest level of the forestry department, directly under the county forestry bureau or township government) provided other necessary services, such as farmer training, extension services, seedling procurement, etc.

Here, government provided a series of services, including planning, capital investment and technical support. Without planning by government, it is quite difficult, if not impossible, to organize such a large-scale plantation. Capital investment and technical skills are often lacking among farmer households, as well. This may explain why the plantation was established only after the government was involved, rather than as a result of household management policies alone. Therefore household management is not a sufficient condition for the establishment of large plantations.

---

Local government played a positive role in Shuiwei, in part because of the kinds of personal networks between officials and villagers discussed above. Mr Jiang Xinglong, deputy director of Libo County Bureau of Forestry, for example, played an important role in organizing the development of a Chinese fir plantation in Shuiwei, partly because of his personal ties to the area. There are also financial incentives for local government to support farmers' efforts to manage forests. Following fiscal decentralization during the 1980s, local governments collect revenues from local economic activities, including taxes and fees on timber and taxes on non-timber forest products, such as bamboo shoots and resin.

However, this fiscal incentive can work to hurt farmers. Obligated to balance their own budgets at the local level, local governments in remote mountainous areas have sometimes overexploited local forest resources and have competed with villagers over shares of the benefits from forest exploit-ation. Local governments have also exacerbated local inequalities. In Lijiang County, for example, Dev Nathan found that the government's Senlong Forest Products Company handled 90 per cent of the timber sales in the county in 1997, and made a net profit of 25 million yuan. During that year, 50 per cent of the net profit of the county-owned company was used for local infrastructure development and other public services. The remaining 50 per cent was invested in forest production. Forest revenues have thus been used to build up the tourist industry in Lijiang City. Naxi people dominate this industry, though much of the timber harvesting occurs in forests where the Yi people are the majority.

Perhaps most importantly, localization does not always work in the interests of forest users because officials are appointed from above, ultimately limiting their accountability to villagers. When opportunities for big pay-offs arise, local officials may be willing to risk the ire of neighbours and forego any financial support from the local community. In Gengma, for example, collective land has been leased by administrative village leaders to investors from the main county or prefecture towns for development as plantations of rubber or fruit trees. Little of the lease money was invested locally, according to farmers, and there was a strong suspicion that the money was simply taken by leaders of the administrative village. It was estimated that, in the process, about 370 hectares of wastelands were leased. Farmers lost access to collective forest that had been a source of many products, including food, fuelwood and medicinal plants.

Where local governments – through personal networks, fiscal incentives or pressure from villagers – have acted in the interest of local forest man-agement, devolution policies have been more successful. Where these same governments have not had any obligation to local farmers, they have contributed to forest degradation and hardship for forest users.

## Issues of policy implementation

The impacts of devolution policies are also determined by how they are implemented, including whether the procedures and guidelines for implem-entation are appropriate and whether officials are accountable for following them. In many cases, the distribution of family plots and responsibility hills to households during the early 1980s was terribly disorganized. In each county, hundreds of government officials and secondary school graduates were seconded to work on the distribution of family plots and responsibility hills. These people had no skills or experience in forestry at all. Moreover, the government demanded that the task be completed in a short time frame (usually a couple of months). In the study villages in Chuxiong, the officials responsible for the distribution of forest land did not seriously mark the boundaries of family plots. Similarly, in Gengma, officials made boundaries on maps at home rather than on the spot in the field. This has led to serious boundary conflicts in both locations today.

Forest fragmentation in family plots and responsibility hills (discussed earlier) increased costs and operational difficulties for households, limiting their benefits from the forestry reforms. It is noted here that the policy did not require the division of forest into many tiny, fragmented plots. This was a decision taken by local officials, one that has slowed the development of local forest management until very recently, with an accompanying rise in shareholding and the leasing and sale of living stands.

## Historical factors

Many foresters and academics blamed devolution policies for the deforestation that took place immediately after the introduction of household-based management during the early 1980s. Based on evidence from our study sites and other sources (Liu, D, 2001), however, the underlying cause of deforestation was a lack of confidence in forest tenure security among villagers. This was the result of frequent changes in forest tenure, especially tenure of private trees, over the period of the mid 1950s to the 1970s. The major changes are as follows:

- In 1956, all private forests were collectivized, with the exception of scattered trees around homesteads (retained by farmer households).
- In 1958, private trees were collectivized, too.
- In 1961 to 1962, collectivized private trees were returned to farmer households from the commune.
- In 1966 and afterwards, private trees were taken away and returned several times across much of China.

After so many changes in tenure policy and practice, farmers were not certain that their tree tenure was secure. It is a logical reaction for them to decide to harvest trees as quickly as they can before the government takes back control over forests once more. Deforestation was the cost Chinese people had to pay for the radical and frequent tenure transformations during the 1950s and 1960s.

## CONCLUSIONS

Devolution is not a panacea for China's woes in managing its non-state forest; but it did help to create and expand space for local forest management. Villages and farmer households obtained more authority to manage forests and to harvest and market forest products. They benefited from forests at greater rates and in more ways, and this gave them incentives to improve their forest management. All of these factors contributed to an increase in forest area and stock, in farmer assets and in forest management capabilities among villagers.

Nevertheless, there are a number of important issues related to devolution's impacts that must be addressed before the policies should be considered a success. Farmers have not always been able to benefit from their newly won tenure rights because of tax and regulatory policies. A lack of capital and technical and marketing expertise has also hampered farmers. Social inequalities and a lack of government accountability have also turned devolution policies against their intended beneficiaries, helping local bureaucrats and wealthy plantation developers to get ahead while the poor lose access to important forest resources. This is especially true of forest leases and contracts.

Fortunately, there are clear strategies for dealing with these issues. Firstly, it is necessary for the government to transfer the rights to harvest and market forest products to communities and farmers. Current restrictions on such rights have severely limited the practical effects of devolution policies by limiting the kinds of benefits that farmers can expect from the trees they now own. Government policies can also be improved by including provisions for more technical extension, improved credit services and market information. More accountable local government and more carefully targeted services will limit the kinds of inequalities that are beginning to emerge in rural China.

The increased activity of Chinese villagers – often working in alliance with one another and with sympathetic government officials – to bring about such reforms is encouraging. So, too, are the examples of local initiatives in forest management, such as the self-initiated shareholding systems, that indicate where farmers would like to take forest management. There is reason for optimism in China as a government sensitive to the needs of its farmers responds to their recommendations for reform.

# 3

# Devolution as a Threat to Democratic Decision-making in Forestry? Findings from Three States in India

*Madhu Sarin, Neera M Singh, Nandini Sundar and Ranu K Bhogal*

## INTRODUCTION

This chapter looks at two trends that are shaping the devolution of forest management in India:

1   the *appropriation* of space for forest management by diverse, self-initiated community formations at the grassroots level; and
2   state-driven *devolution*, where government policies define the scope of local authority in forest management.

We assess whether state devolution policies are increasing or decreasing local forest users' space for exercising democratic local control over forest management decisions, enhancing livelihoods and improving forest quality.

We focus on two devolution policies currently relevant in India: Joint Forest Management (JFM), which solicits local peoples' 'participation' in state forest management, and the Panchayats (Extension to the Scheduled Areas) Act, 1996 (PESA),[1] which devolves considerable authority to self-defined and self-organized communities to manage their local forest resources. Local may include single or multiple villages, or hamlets within them, or sub-groups of youth or women within any of the former. Institutional structures may include formally constituted *gram panchayats* – the lowest level of local self-government, self-organized or traditional village institutions; mobilized *gram sabhas* – the collective body of adult electors of self-defined communities under PESA; or women's or youth associations, or government-promoted participatory village 'committees'.

Studies were undertaken in three states: Orissa, Madhya Pradesh (MP) and the Uttarakhand region of Uttar Pradesh (UP).[2] All three states have ongoing JFM programmes, the latter two funded by World Bank loans. They also have

various pre-existing community-based forest management (CBFM) institutions, although only the legally demarcated village forests managed by
elected *van panchayats* (forest councils) in Uttarakhand are officially recognized. In contrast to the more developed regions of the country, the study sites
have 38 to 67 per cent of their total geographical areas – representing erstwhile
uncultivated commons – that the government has declared as state-owned
forest land. Large proportions of their population live below the poverty line
and forest lands claimed by the state continue to be critical for supporting local
livelihoods. Consequently, all three are sites of significant 'appropriated' space
for local forest management. All three regions, as most other forested regions
in the country, have histories of local rebellions and struggles for protecting
community rights against forest reservation, ruthless commercial exploitation
and 'development'-induced displacement by the state, both during the colonial
period and since independence.[3]

These provide valuable insights into community perspectives on JFM and
other forms of state-driven devolution. They indicate where state and local
interests in local management diverge, how the two interests affect and
reshape one another, and the outcomes of such interaction for local people and
forests. Analysis of these three states highlights the changes required for
making state-driven devolution policies more responsive to the priorities of
forest-dependent women and men. Forest-dependent people have developed
their own strategies for shaping such policy reform. These include building
alliances between CBFM groups through federations in Orissa (Vasundhara,
2000), asserting community interpretations of the empowering provisions of
PESA in pockets of MP (Behar and Bhogal, 2000; Sundar, 2000b), and challenging
the parameters of existing devolution policies and donor-funded projects
through advocacy and people's movements in all three states (Vasundhara,
1999; Behar and Bhogal, 2000; Sundar, 2000a; Sarin, 2001a, 2001b and 2001c).

## THE HISTORICAL CONTEXT FOR DEVOLUTION POLICIES

The history of forest control and management in India is important for
understanding the continuum of interactions between the state and local
people, and the hazards of taking the political, economic and environmental
claims of the state about present devolution policies at face value. *Set in a larger
policy and historical context, devolution policies emerge as a further extension of state
control, at best a meager palliative for mobilized forest users, rather than as a real move
towards greater democracy, improved local livelihoods and healthier forests.*

State appropriation of the uncultivated commons, termed 'the wastes' by
the colonial government due to their not yielding land revenue, began during
the late 19th century. The 1878 Indian Forest Act, precursor to the present
Indian Forest Act (IFA) of 1927, created three classes of forest – reserve, in
which people have no rights unless specifically recorded; protected, with all

rights unless specifically forbidden; and village forests for meeting local needs. Although areas for meeting community needs were set aside in response to protests and rebellions against forest reservation, none of these were declared village forests under the IFA during the colonial period. Forest classification fragmented people's holistic livelihood base into different legal categories. Access to reserve and protected forests was further fragmented through the allocation of individual rights, privileges or concessions administered by the state.

*Reservation of forests under these acts was probably the single most important turning point in forest-people relations in rural India.* The forest acts were powerful legal instruments. In areas such as Uttarakhand, forest reservation was accompanied by forest 'settlements' involving the recording of customary rights of users. In other areas, such as Bastar in MP, some people were physically removed from forests, while for others twice the area of cultivated land was left aside for villagers' domestic use (Sunder, 2000). In many cases, sweeping notifications were issued declaring all uncultivated 'wastes' as protected forests. The legal designation of such lands as forests has remained frozen until today, irrespective of whether they have, or ever had, forest cover or not.

Post-independence, the Indian state continued appropriating the commons.[4] Between 1951 and 1988, the *net* area under the control of forest departments increased by 26 million hectares (from 41 to 67 million hectares), predominantly as reserve forests (Saxena 1995a, 1999) in which people have limited or no rights. Ironically, large parts of this consisted of areas set aside for community use under local management during the colonial period. In the tribal district of Bastar in MP, for example, all *nistari* (community) forests, left alone by the colonial state in response to violent rebellions by the *adivasis* (tribal people),[5] were declared state protected forests in 1949, just two years after independence. In Orissa, also, large areas of the commons under the jurisdiction of former princely states or large landlords, in which the villagers had extensive customary rights, were 'deemed to be (state) Protected Forests' under the IFA (GoO, 1996, p28) during the early 1950s.

The state also centralized forest administration after independence. In 1976, forests under the jurisdiction of individual states were moved to the concurrent list of the constitution, empowering the Indian government to have a decisive say in forest management priorities. The Forest Conservation Act (FCA) of 1980 made central government permission mandatory for converting even small parcels of forest land to non-forest uses. Centralization severely reduced villagers' access to basic development facilities in forested regions and prevented state governments from meeting the basic needs of their rural populations without central government clearance. The *Chipko* (hug the trees) Movement of the 1970s was, in part, a response to the indiscriminate commercial felling by the state in Uttarakhand. Yet, activists of the same movement initiated a Tree Felling Movement during the late 1980s in protest against the interminable delays in obtaining central government's permission to use forest land for basic development facilities for hill villages. Until 1980, the state governments periodically granted legal tenure to *de facto* ancestral cultivators and settlers. Following the FCA, they can no longer legalize supposedly illegal forest 'encroachment' without central government approval. In May 2002, the Central Ministry of Environment and Forests issued a circular to all state governments to evict all encroachers on forest land by the end of September. An estimated 10 million tribal forest dwellers faced possible eviction from their ancestral lands simply because, in many states, they were never provided land title deeds due to their lands being declared state forests without comprehensive surveys.

With increasing market value of what was originally classified minor forest produce (MFP), from the 1960s onwards, important non-timber forest products (NTFPs) were also nationalized, and monopoly control over their collection and marketing was vested in state forest corporations or other agencies. During the 1960s and 1970s, many states also appropriated ownership or monopoly marketing rights over the more valuable NTFPs and tree species, even from private lands.

During the past three decades, state appropriation of forest resources has been increasingly justified in terms of conservation goals. The national Wild Life Protection Act, 1972, (WPA) enables the physical displacement and exclusion of local villagers from protected areas. Over 4.5 per cent of India's geographical area has already been brought under the protected area network by a decision-making process in which the people living in, or dependent upon, these areas have had no role. This has led to acute conflicts between the affected villagers and the protected area managers. The geographic distribution of biodiversity and wildlife-rich areas within the country overlaps with areas that have the highest concentrations of tribal people, underdevelopment and poverty. The livelihood and human rights of such already marginalized people have been inequitably appropriated for providing ecological and conservation benefits to distant regional, national and global interests.

To sum up, increasing state control over a century has de-linked forest-dependent communities from managing local forest and land resources. The state has broken the uncultivated commons into different legal categories with differentiated access for different categories of people. State interventions have further atomized the resource into specific products and services, and converted holistic local authority into a series of rights, concessions and privileges, granted at the pleasure of the state. The state has monopolized the right for commercial forest exploitation and conservation. Though met with frequent and widespread resistance, this process has fundamentally restructured people–forest relations.

## A change of heart or a change in tactics?

### New forest policy and Joint Forest Management (JFM)

Increasing prominence of environmental concerns and the human and resource rights of indigenous forest-dwelling communities from the 1970s led to major reversals in forest management priorities during the 1980s. In contrast to the earlier focus on maximizing revenue and promoting forest-based industry in national interests, the new 1988 forest policy of India articulated the twin objectives of ecological stability and social justice. Highlighting the symbiotic relationship between tribal people and other poor people and the forests, the new policy emphasizes protection of their rights and treating local needs as 'the first charge' on forest produce (GoI, 1988). Stating the need to generate 'a massive people's movement, with the involvement of women' for achieving its objectives, for the first time, the 1988 forest policy created space for the participation of forest-dependent women and men in the management of state-appropriated forest lands.

The ambiguities in the dual objectives of the 1988 policy, however, have left considerable room for the politics of interpretation. While social activists emphasize the policy's support for forest dwellers' rights and improved livelihoods, conservation interests highlight its environmental and ecological objectives. Although, in principle, the two objectives can be met simultaneously, they have often conflicted, in practice. The social justice objectives have had no clear legislative support, whereas the environmental objectives have been enforceable with the Forest Conservation and Wild Life Protection Acts. Livelihood activists have been at a clear legal disadvantage in confrontations with conservation interests.

For translating the participatory policy objectives into practice, the Central Ministry of Environment and Forests (MOEF), issued a circular on 1 June 1990 to all states and union territories providing guidelines for the 'Involvement of Village Communities and Voluntary Agencies in the Regeneration of Degraded Forests' (GoI, 1990). This led to the adoption of what has come to be called joint forest management (JFM) by several state forest departments. As of October 2001, 27 of 28 states had issued JFM orders, specifying their respective terms for working in partnership with local villagers, and 14.25 million hectares of forest land (18 per cent of total forest area) were already officially protected by roughly 62,890 village organizations under JFM (INFORM, 2001).

In February 2000, the MOEF issued revised guidelines for JFM. These permitted a cautious extension of JFM to well-stocked, instead of only degraded, forests; specified women's representation in JFM groups (a minimum of 33 per cent in executive committees and 50 per cent in the general body); and clarified that JFM micro-plans[6] must conform to the silvicultural prescriptions in the forest departments' technical working plans. The new guidelines suggested that all village organizations participating in JFM should be registered as societies in order to provide them with an independent legal identity, but remained silent on assuring them tenurial security over JFM land. The new guidelines also recommended that all such organizations should be called 'JFM committees' across the entire country, irrespective of their diverse histories, legal status and institutional structures (MOEF, 2000). State JFM orders vary in terms of the legal status of the land to which JFM may be extended, and the organizational structure, autonomy and entitlements of village institutions participating in JFM. Some states have restricted JFM to legally notified degraded forest lands, whereas others, including Orissa and UP, have extended it to revenue lands under revenue department jurisdiction. The JFM orders of Rajasthan and Karnataka brought even village grazing and other common lands under JFM's ambit. In Uttarakhand, JFM brought the only autonomously managed community forests with legal standing under joint management with the forest department. Present devolution policies are effectively continuing historical trends of appropriation by bringing remaining common lands under JFM. *The unexamined assumptions are, firstly, that their most appropriate use is forestry, irrespective of diverse existing uses by local communities, and, secondly, that the forest department is the most competent manager for them.*

## Decentralization of governance and PESA

A related, significant shift to decentralization was brought about by the 73rd amendment to the Indian constitution in 1992 that made it mandatory for all states to decentralize governance through a three-tier structure of Panchayati Raj (local self-government) institutions. The 29 functions recommended for decentralization to Panchayati Raj institutions include common lands, social forestry, fuelwood plantations and non-timber forest products (NTFPs). Management of nationalized forest lands is not included unless specifically notified by a state government. The relationship between democratic Panchayati Raj Institutions and parallel village organizations for JFM promoted by the Forest Department has been a major issue of debate and discussion in policy advocacy and analysis forums. Yet, the February 2000 central guidelines for JFM do not even mention the role of local self-government institutions in JFM.

Enactment of the central provisions of the *panchayats* (Extension to the Scheduled Areas) Act (PESA) in 1996 provided a more radical constitutional and legislative mandate for devolution of local self-governance in Schedule V (tribal majority) areas. PESA makes the *gram sabha* (the body of all adult voters of a self-defined community) 'competent to safeguard and preserve the traditions and customs of the people, their cultural identity, community resources and the customary mode of dispute resolution' (GOI, 1996, clause 4d). Every *gram sabha* is also empowered to approve the plans, programmes and projects for its social and economic development before their implementation, besides having ownership of minor forest produce within its area, either directly or through the *gram panchayat*. Most states (including MP and Orissa) with Schedule V areas have enacted their state acts under PESA as stipulated by the central act.

PESA is arguably the most empowering legislation for India's tribal people, supported by an amendment of the Indian constitution. Yet, as in the case of the 1988 forest policy, the act is riddled with ambiguities, making it equally vulnerable to the politics of contradictory interpretations. PESA effectively mandates community-based forest management by *gram sabhas* in Schedule V areas. This is in contrast to JFM, which establishes new village committees under forest department supervision to manage forest lands. Forest department and administrative officers assert that PESA does not apply to the management of nationalized forest lands. Mobilized tribal communities, on the other hand, insist that their community resources include forest areas that they have traditionally used, but which have been taken over by the state. That the revised JFM guidelines do not even mention PESA indicates the low importance given by the central Ministry of Environment and Forests to the bearing of its provisions on the JFM framework in Schedule V areas. PESA co-exists with totally contrary legislation, such as the Land Acquisition, Wild Life Protection, Forest Conservation and the Indian Forest (1927) acts, with no clarity about which is to prevail over the others. Whereas PESA remains largely unimplemented, enforcement of the other acts continues as before. The central government, in fact, has drafted a proposal to amend Schedule V of the constitution itself to open up tribal areas for commercial exploitation by national and multinational corporate interests (Pattnaik, 2001).

It is against this backdrop that we analysed state-driven devolution policies and locally appropriated space for forest management. The following sections discuss both processes in Orissa, MP and Uttarakhand, and examine how local women and men have appropriated space despite government policies. The sections evaluate what sort of spaces for local forest management were created or destroyed by each and analyse why that happened. In Orissa and MP, we compare JFM with self-initiated forest management commonly, referred to as Community Forest Management (CFM). In Uttarakhand, we compare Village Forest Joint Management (VFJM) with *van panchayat* management, as well as with informal CFM. *Unfortunately, as the cases indicate, devolution policies have largely reinforced state control over forest users, giving the relationship new form rather than changing its balance of power or reducing the conflict between state and local interests.*

# ORISSA

Orissa is one of the poorest states in India, with 49 per cent of the rural population living below the official poverty line (GoO, 2000). With scheduled castes and scheduled tribes, respectively, comprising 16 per cent and 22 per cent of the population, and NTFPs estimated to contribute between 15 to 40 per cent of rural household income (Vasundhara, 1998), forest lands are a critical livelihood resource for the rural poor. This dependence has shaped local responses to forest degradation in the form of CFM. Orissa, more than any other state in India, provides extensive examples of the appropriation of space for forest management by local villagers. Local initiatives to protect and manage forests through community-based arrangements have existed on a large scale in Orissa for several decades. During the late 1980s, these self-initiated forest management groups intensified their demand for official recognition and devolution of forest management authority to them. The state response to these demands has been joint forest management.

JFM has limited space for local initiatives in many ways in Orissa. Firstly, villagers' decisions require ratification by the forest department under JFM.

**Table 3.1** *Site characteristics – Orissa*

| Site | Suruguda | Baghamunda | Berham | Jhargaon | Aonlapal |
|---|---|---|---|---|---|
| District | Sundergarh, Schedule V area | Deogarh | Angul | Jharsuguda | Balasore, Schedule V area |
| Nature of space for local forest management | Self-initiated, formalized under JFM | Self-initiated, no JFM | Self-initiated for DPF, JFM for RF | Self-initiated, formalized under JFM | Self-initiated, formalized under JFM |
| Legal status of forest | Reserve forest (RF) and *khesra* forest (KF) | RF, social forestry plantation (VF) | RF, KF converted to demarcated protected forest (DPF) | Proposed RF, KF | RF |
| Forest type and quality | Dry deciduous, sal | Sal dominated | Mixed forest | Mixed forest | Sal |
| Number of village households | 155 | 26 | 199 | 100 | 87 |
| Presence of village institutions prior to CFM | Yes | Yes | Yes | Yes | Yes |
| Prior experience of managing other resource commons | No | No | Yes | Yes | Yes |

| Site | Suruguda | Baghamunda | Berham | Jhargaon | Aonlapal |
|---|---|---|---|---|---|
| Reasons for initiating forest protection | Acute scarcity of forest products and fuelwood due to forest degradation | Acute shortage of wood for agricultural implements and fuel | Degradation of adjoining RF intensified pressure on KF; villagers then decided to protect KF | Shortage of wood for agricultural implements and house construction due to forest degradation | To prevent neighbours from damaging local rice fields on the way to their forests and to stop forest degradation and scarcity of forest products |
| Year of initiation of forest protection | 1985 | 1991 | 1990 | 1970–1971 | 1980 |
| Dependence on forests | Medium | High | High | Medium | High |

*Notes:*
RF: reserve forest (under forest department)
KF: *khesra* forest (undemarcated protected forest under revenue department)
VF: social forestry plantation, notified as a village forest (mostly under revenue department)
DPF: demarcated protected forest (under revenue department)

Secondly, local institutions that are adaptable to changing social and environmental conditions are replaced by an institutional arrangement uniformly
applied. Thirdly, forest benefits under JFM need to be shared with the state in
accordance with a rigid, unilaterally defined formula that does not take diverse
local livelihood needs and forest values into account. And fourthly, technical
management has to conform to the forest department's vision of a good forest.
Instead of any real devolution of authority, the state has manoeuvred devolution policies to regain control over local forest management initiatives by
imposing its own institutional arrangements on them. Forest-dependent
communities have, in turn, responded by attempting to preserve local control
both within and outside the context of JFM. They have continued to lobby for
greater community-based rights over forests and forest products using a
variety of advocacy tools and processes, including federating at various levels
in order to strengthen CFM.

We present our methods in the section below and then contrast the impacts
of CFM and JFM. We conclude by looking at how local forest-dependent
groups are responding to JFM and discuss their evolving strategies and
dilemmas.

## Methods and site selection

Five participatory village case studies were selected to demonstrate a range
of community initiatives and their interface with JFM. Policy constraints and
local responses were also analysed for six NTFPs. Process documentation of
advocacy approaches and developing federations of CFM groups included
interviews with key advocacy groups, federation leaders and forest department field staff.[7] Archival research on forest management and administration
in several princely states that merged to form the state of Orissa attempted to
place present devolution policies in their historical context. Characteristics of
the five study sites are summarized in Table 3.1.

## Community Forest Management (CFM) in Orissa

### Origins of CFM

In the words of the women of Baghamunda:

> We realized that if we do not protect our forest, it would be gone; and we
> would be the ones most affected by that . . . the forester would be least
> affected. He will still get his salary. For us, nothing would be left.

Officially, about 37 per cent of the total geographical area of Orissa, or about
58,000 square kilometres, is recorded as state-owned forests. Of this total, 45.3
per cent consists of reserve forests under the administrative jurisdiction of the
forest department. Another 26.7 per cent consists of demarcated protected
forests (DPFs) and undemarcated protected forests (UPFs), and the remaining

28 per cent of village and unclassed forests, under the jurisdiction of the revenue department, also referred to as revenue forests (GoO, 2001).

Official data about state forest land, however, needs to be treated with caution since the settlement of rights and the demarcation of boundaries have either not been done at all or suffer from serious infirmities. 8.8 per cent of the state's total forest area is under shifting cultivation (FSI, 2000) with poor, often non-existent, recording of the tribal shifting cultivators' ancestral rights. According to recent data, only 30 per cent of the state's total area had any forest cover and only 16.74 per cent had dense forests (FSI, 2000). Common revenue lands under diverse administrative, managerial and rights regimes in the erstwhile feudatory states were 'deemed to be protected forests' after merger of the feudatory states with the Indian Union after independence. There are no detailed surveys or management plans for these lands, and there is ambiguity over their line of control. While they come under the jurisdiction of the revenue department, responsibility and control over their technical management is vested in the forest department, which leads to confusion.

According to latest available estimates, over 5000 out of 12,000 villages within or near forests (out of approximately 51,000 villages in Orissa) are actively protecting and managing state-owned forest lands in their vicinity.[8] While some villages have been protecting their forests for as long as 60 to 70 years, the majority started protection during the late 1970s or early 1980s. The management unit ranges from a group of households, to a settlement or hamlet, to a cluster of villages, while the area under protection ranges from a only few to 1000 or more hectares. The revenue forests are generally in small patches (50 to 1000 hectares), interspersed among agricultural fields and surrounded by several villages. Under many of the feudatory states, parts or all of these areas were meant for the villagers' use as *gramya jungles* (village forests). Villagers' substantial customary rights in these forests, combined with the weak presence of the forest department in them, facilitated community initiatives.

Forest dependence was moderate to high in the five case study sites (see Table 3.1), and villagers often commented that their motivation for initiating forest protection was the loss of an important forest product or service. In many cases, the initiative to protect forests came from the farming community, primarily out of an interest in ecological services, such as reducing soil erosion and increasing water retention in soils. Scarcity of forest products and wood for house construction and agricultural implements was an especially significant factor for the poor. CFM was most pronounced in areas that had witnessed rapid forest degradation over a short period, with drastic impacts on people's lives. When some villages start restricting outsiders' access, other villages are prompted to enclose their own patches, as well.

Various organizations evolved within villages and hamlets or between villages. These include forest protection committees with an executive committee selected/elected by the village general body, usually excluding women. Other organizations include councils of elders, youth clubs or, occasionally, *mahila samities* (women's organizations). Villagers generally start with a few simple rules that gain in complexity with changing situations. Adaptation also takes place in group formation and composition. Groups may dissolve or reform based on dissatisfaction with earlier arrangements. Protection systems varied between *thengapalli* (voluntary patrolling in rotation) and paid watchmen. CFM groups often had elaborate rules for penalizing those from within and outside of the community. Enforcement of these rules was performed through social sanctions in the absence of formal legal authority. These arrangements are dynamic and adaptive.

In all the study sites, and as corroborated by several earlier studies (Kant et al, 1991; Jonsson and Rai, 1994; Singh and Singh, 1993, 1994; Singh, 1995; Vasundhara, 1997, 1999; Sarin and Rai, 1998; Conroy et al, 2000), the presence of village institutions and their prior experience with managing common property resources was a major factor in contributing to the emergence of CFM. Although most of the commons have been appropriated by the state, collective community-based arrangements still exist for managing water bodies and village *haats* (markets), celebrating cultural and religious events[9] and managing conflicts. The weak presence of Panchayati Raj institutions for local governance – due to infrequent elections until 1993 – also contributed to the development of informal village institutions. All five sites had vibrant village institutions prior to forest protection. In three cases, the villagers had prior experience of managing other common pool resources. In Jhargaon and Berham, the villagers had well-defined arrangements for managing common lands used for temple upkeep, water-harvesting structures and springs used for irrigation.

## CFM: Regeneration of forests and local institutional capacities

CFM has had dramatic, positive impacts on forest quality, as well as on enhancing the capacity of local institutions to deal with issues relating to villagers' lives and livelihoods. Most villages began by protecting forests that had been extremely degraded. In many cases, the villagers had even started digging up tree roots for cooking fuel. In all our case study sites, both the sal-dominated and mixed forests had responded very well to protection. Within a period of ten years, they had regenerated substantially to meet local requirements of forest products, as well as improving water availability and soil fertility. In all the cases studied, the villagers were highly enthusiastic about their regenerated forests. In Suruguda Village, char seeds were more plentiful and provided an important source of income. In Jhargaon Village, regenerating medicinal plants were meeting the requirements of local traditional health practitioners. In Aonlapal Village, sal leaves are now available throughout the year, and provide sufficient material for households engaged in leaf-plate stitching (see Box 3.1). In Jhargaon and Baghamunda villages, wildlife returned with forest regeneration. In Baghamunda, forest regeneration reduced soil erosion and improved water regimes for agricultural lands in lower catchment areas. The villagers took up regular thinning and pruning operations, using the wood as fuel. Villagers also got a variety of edible tubers, leafy vegetables, fruits, mushrooms and other products for consumption and sale in local markets.

---

**Box 3.1** *Sal-leaf plate stitching*

Sal-leaf plate stitching is a significant source of income in Aonlapal Village. Out of 83 households, 12 are landless and earn 8000–9000 rupees per year from stitching. Another 33 households take up seasonal leaf stitching to supplement agricultural income, earning 4000–5000 rupees per household over a five- to six-month period. Some members of landless households who had migrated to cities for employment returned to the village upon improvement in sal leaf availability following forest regeneration. Recent moves by the forest department to clamp down on the sal leaf trade on the grounds of sustainability, however, are likely to adversely affect the livelihoods of poor villagers dependent upon this trade.

---

Many CFM villages in Orissa also use their income from forests, mostly from occasional cleaning and thinning operations, for building village assets. Among our case study villages, the salaries of two teachers at the upper middle school were paid from the forest fund in Jhargaon Village. Suruguda Village had received 50,000 rupees in 1989 as a national award for its exemplary environmental conservation work.[10] Some of this money was used to construct a building for the Forest Protection Committee (FPC) and for repairing the village school; the rest was kept as a fixed deposit in the bank for future use. In Berham, forest income was used to provide loans to needy persons at low interest rates.

In all the case study villages, forest regeneration improved the availability of NTFPs and related incomes for poorer women and men, although inequity in the distribution of benefits existed. Distorted NTFP markets and lack of marketing support by the government still left NTFP gatherers at the mercy of petty traders and middlemen. Lack of clear tenurial rights further constrained local communities in tapping the full potential of forests for supporting local livelihoods and village development. Nevertheless, forest regeneration restored some of the place of NTFPs in the household economy of the gatherers.

Collective action for forest protection also strengthened local institutions, enabling villagers to take up management of other common pool resources. In Baghamunda, women's involvement in forest protection increased their self-confidence and ability to deal with the outside world, including government officials; the women's association is now playing a leading role in the campaign for higher prices for *kendu* (*Diospyros melanoxylon*) leaf. Collective action at the village level also led to inter-village cooperation and the development of federations of CFM groups, which are playing an important role in addressing livelihood concerns. The district forestry forum in Balangir, for example, successfully lobbied for better prices for NTFPs, support for marketing of surplus agricultural commodities, as well as for timely availability of seeds. In Ranapur, the block-level federation was successful in thwarting a state move to privatize the management of cashew plantations on revenue lands. In Dhani Panch Mausa, the FPC was able to access government funds for a leaf-plate stitching machine for a women's group, as well as dairying loans for the poorest village households, providing them with alternative livelihoods to fuelwood headloading.

CFM has created substantial space for local forest management in Orissa. Based on local demand and building upon existing local institutional forms and management practices, CFM has improved forest quality, supported local livelihoods and improved local institutional capacities. However, as is common elsewhere (Sarin et al, 1998), dominant sections within villages have benefited disproportionately from the regenerated forest, and have been able to impose their perspectives on how forests should be protected and utilized upon the poorer and marginalized sections. We discuss such equity issues in greater detail in the section 'Equity concerns in CFM' below.

## State response to CFM: from a 'conspiracy of silence' to JFM

The initial state response to these CFM initiatives was to ignore their existence. A social forestry project, implemented in Orissa during the 1980s, is illustrative. The project spent large amounts of money on village woodlot plantations on *non-forest* land in order to 'save' the forests from the very people who were protecting them. Towards the end of the social forestry project, a component for supporting regeneration of degraded sal forests through community involvement was introduced. Foresters, however, continued to show little interest in this work. According to a range forest officer, most social forestry project deputy directors preferred to surrender their natural regeneration targets to compete for increased targets for village woodlots because these had higher budgetary allocations. Controlling larger budgets, rather than adopting cost-effective, community-based methods for improving forest quality, was the evident incentive for Forest Department staff.

With the social forestry project ending in the mid-1990s, the Forest Department's attention shifted back to natural forests. By then, community initiatives were too visible to be ignored.

From the mid-1980s, CFM groups started pressuring the state government to grant them formal recognition. During 1986 to 1988, villagers in Orissa, with the support of social activists, academics and NGOs, conducted a state-wide postcard campaign addressed to the state chief minister, demanding rights over community-protected forests. In response, a government resolution was issued by the Forests, Fisheries and Animal Husbandry Department (the Forest Department, for short) in 1988 that provided for villagers' involvement in the protection of reserve forests. This, however, did not offer them much. They were only permitted to meet their *bona fide* domestic requirements of fuel and small timber free of royalty. A revised resolution in 1990 extended this provision to protected (both demarcated and undemarcated) forests falling under the jurisdiction of the revenue department. In 1993, the forest department issued yet another government resolution – more in line with the standard JFM framework that was developed in other states during the intervening period – as a further response to mounting grassroots demands for greater community forest rights. The 1993 JFM resolution provided for sharing 50 per cent of the income from timber and access to NTFPs (from both reserve and protected forests) with FPC, called *vana samrakhana samitis*, to be constituted by the forest department.

In 1996, the forest department issued yet another government resolution that, in some respects, could be termed revolutionary in the present Indian context. This allowed forest areas under villagers' protection to be declared village forests by the forest department, granting villagers rights to manage all NTFPs within them. The 1996 resolution was issued in response to the then chief minister's insistence on giving greater forest rights to local villagers. This resolution, however, has remained a dead letter. The Orissa Ministry of Forests and the Environment (under which the forest department falls) issued it in haste, without consulting the foresters responsible for its implementation, in order to satisfy the then chief minister. The foresters, for their part, effectively subverted the policy by not implementing it.

A range officer's comment on the 1996 resolution reflected the forest department's attitude. According to him: 'How can you possibly declare reserve forests as village forests? Reserve forests are of much "higher status". How can you expect a college or graduate school student to go back and study in primary school'! The forest department argued that the Forest Conservation Act (FCA) prohibits the de-reservation of reserve forests. Section 28 of the Indian Forest Act (IFA), however, specifically permits the state government to designate any reserve forest as a village forest, to be managed by the community. A committee was set up to recommend how best to operationalize the 1996 resolution and took a year to give its recommendations. These included that:

- a beginning should be made by declaring protected forests as village forests, while seeking legal opinion on converting reserve forests to village forests;
- at least five villages per forest division should be declared as village forests on a pilot basis; and
- forestry field staff should be given clear operational guidelines for implementation.

No action was taken on any of the committee's recommendations.[11] The 1993 JFM order has continued to be the operational one.

Moreover, as revenue forests are under the jurisdiction of the revenue department, and the undemarcated ones among them are only 'deemed to be protected forests', attempts by the forest department (which is not the landowner) to extend JFM to them (and to claim 50 per cent of the income from them in the process) are of dubious legality (Mishra, 1998). Further complicating the situation, the Orissa *Gram Panchayat* Act, 1965, vested the management of *gramya jungles* (coming within the fold of protected forests) within revenue village boundaries with the *grama panchayat*. The 73rd constitutional amendment also mandated devolution of such resource management resp-onsibilities to the *gram panchayats*. In Schedule V (tribal majority) areas, PESA further devolved natural resource management authority to the *gram sabha*. All of these legal provisions have been ignored by the forest and revenue departments and even by the *gram panchayats*, with most foresters and *panchayats* unaware of them. A recent study of legal aspects of JFM and CFM (Mishra, 1998) pointed out that the forest department does not have any *locus standii* for negotiating JFM partnerships with villagers for protected forests. There is thus utter jurisdictional confusion over 55 per cent of 'forest' lands in the protected and unclassed forest categories in Orissa, as well as the legal status of both JFM and CFM on them.

Thus, by 1993, local forest management initiatives had encountered state-driven devolution policies in a struggle over the nature of local people's role in state forest management. Although the state was forced to recognize community forest protection efforts, it has attempted to regain control by recognizing them on its own terms.

## Tensions between the expectations of CFM groups and JFM[12]

The imposition of JFM on locally initiated CFM created tensions that made forest management more difficult, threatening the well-being of forests and forest users, alike. The major sources of tensions are discussed below.

### *Benefit-sharing*

Orissa's 1993 JFM resolution provides a 50 per cent share in major/final harvest and 100 per cent of intermediate produce to the *vana samrakhan samitis*, the local organization created by the forest department. A general response of CFM groups to this was: 'Where was the forest department all these years when the villagers were struggling with forest protection on their own? Why has it now come to claim a 50 per cent share?' The villagers feel that instead of 'giving' them 50 per cent, JFM actually takes away 50 per cent of what is rightfully theirs.

Similarly, 100 per cent of the intermediate produce (including NTFPs) from JFM forests is supposed to go to the *vana samrakhan samitis*. However, until 1999, monopoly collection and marketing rights over 29 NTFPs from all forests were leased to a private trader. The state-fixed NTFP prices were paid to col-lectors on the basis of minimum wages for unskilled work and not on the basis of the value of the produce. Thus, even on supposedly 'jointly' managed forest lands, the co-managers were treated as mere labourers. Policy changes in the year 2000 maintained state control over valuable NTFPs, transferring control over 67 low-value products to *gram panchayats* under Orissa's Gram Panchayat Act. Until the end of 2001, even this transfer of control was notional as *gram panchayats* were granted only regulatory powers, and no efforts were made to improve the bargaining power of NTFP gatherers.

In several cases, the Orissa Forest Department also did not honour the commitment to share 50 per cent of the bamboo harvested from JFM forests with the villagers. Bamboo forests under JFM were often leased to paper industries that harvested them without consulting or sharing any produce with the villagers (see Box 3.3).

## Differences in forest values and forest management priorities

There was considerable difference in the perceptions of forest-dependent women and men and the forest department regarding the value of forests and management priorities. Many village leaders felt that the forest department's management objectives did not change under JFM. Benefit-sharing under JFM reflected a continuing state focus on revenue and timber. CFM groups preferred a focus on location- and group-specific fulfillment of their needs, especially flows of NTFPs and ecological services from forests. During a discussion on benefit-sharing under JFM at a meeting in Balangir, an old man walked out of the meeting saying: 'We have nurtured the forest as our child; now you are discussing who would take what percentage of it!' In Nayagarh District, the district-level federation leaders emphasized: 'Our forest is not a crop. Do not impose your share-cropping model on us.' This sentiment was echoed in many places and in many forms (Vasundhara, 1999). In the absence of effective participation of village women and men in *how* forests should be managed, sharing products or profits with them is totally inadequate.

## JFM's weak legal standing

Forest protection groups, with support from NGOs, advocated clear community rights over forests and forest products. Villagers felt that these rights were theirs by virtue of their protection of regenerating forests, by their dependence on forests for livelihoods, and by their often long-standing presence and customary forest rights. They also argued that clear legal and tenurial rights were necessary for enabling local communities to realize the full benefits from their self-initiated management efforts. JFM, however, did not offer tenurial security. The JFM resolutions of most states remain mere administrative orders, unsupported by changes in forestry legislation. The colonial IFA of 1927 continues to provide the overarching legal framework for forest management and administration in the country.

During village, village-cluster and district-level workshops facilitated by Vasundhara and another NGO (Sanhati), villagers repeatedly raised the problem of weak community forest tenure under the existing legal framework. This did not mean, however, that everyone advocated full community ownership of forests. Instead, a number of alternative rights regimes that would better meet local needs were articulated (see Box 3.2).

---

**Box 3.2** *Alternative community rights regime for forests in Orissa*

While there was a general consensus among villagers on the need for local rights and control over managing forests as a community resource, there were disagreements regarding the nature of community control and the mechanics of exercising it. Elite male leaders generally argued in favour of a complete transfer of forest ownership to individual groups defined by specific geographic boundaries. Others, particularly women, opposed this on the grounds that those dependent on forests came from larger spatial landscapes whose rights also needed protection. 'Community' ownership can imply strong rights of exclusion and transformation, often at the expense of forest-dependent people. A community rights regime must address the access of non-timber forest product (NTFP) gatherers (most of whom are impoverished women) from villages without their own forests to forests claimed by other villages. In a discussion on community property rights, Sumoni Jodha, a leader from Ama Sangathan, an apex organization of women's associations based at Kashipur, said:

> We don't believe in complete rights over the forest being given to individual communities or villages. Women from villages who do not have forests also come to our forests to collect broom grass. That cannot and should not stop. These women are also poor and need to collect various forest products for their survival. We cannot engage our 'forest guards' and stop them from coming to 'our' forests.

In many CFM areas, restrictions on access to valuable NTFPs have already been imposed on outsiders, even those with a history of access and current dependence on the forest. While regulation of access is needed to check overexploitation, there is a strong danger of control over forest resources being concentrated in the hands of an elite minority, given the stratification and inequalities in village society.

CFM requires a shift in *who* is managing forests and *how* forests are managed. The challenge lies in developing self-governance systems based on principles of participatory democracy and ethical values of equity, and creating conducive conditions for informed and gender-equal collective choice arrangements.

---

## Locus of decision-making

In all Indian states, JFM structures an imbalanced power relationship between the forest department and local communities, with the department retaining control over most forest management decisions. Forest protection groups in Orissa are unwilling to accept such a relationship after having effectively 'controlled' and regenerated their forests. Despite lacking *de jure* autonomy, CFM groups have been taking all forest management decisions *de facto* and are unwilling to see that changed. Villagers particularly object to the superimposition of the *vana samrakhan samitis'* organizational structure, rule-making procedures and bureaucratic culture upon their existing self-governing institutions. They strongly resent the replacement of local leaders by official members, such as the forester (as member secretary), *Naib-Sarpanch* (as president) and ward member (Vasundhara, 1999).

## Forest Department: community relationship under JFM

Despite their reservations about JFM, many villagers considered it strategically important to gain some formal recognition from the Forest Department. Out of the approximately 2000 *vana samrakhan samitis* formed for JFM in Orissa until April 2000, about 1500 were formalized CFM initiatives.[13] Where relations between forestry staff and villages had been good, JFM brought the benefits of Forest Department support for handling forest offenders and providing technical guidance, besides formal recognition. Forest Department support, for example, helped Suruguda to win a national environmental award.

The Forest Department, however, has done little to create confidence that it will provide villagers with the type of support that they expect in most villages. Local resentment was especially strong where villagers failed to get any departmental cooperation when they were struggling to protect their forests. Forest offenders handed over to the Forest Department were often let off, allegedly against petty bribes, or villagers were kept in the dark regarding the action taken in such cases. Demarcation of forest areas between villages has also emerged as a particularly serious problem as most inter-village conflicts are over boundaries. The villagers expected the Forest Department to facilitate boundary demarcation and to give recognition to their own efforts; but little progress has been made thus far. Against such a background of poor accountability and support from the Forest Department, JFM is seen as an intrusion on local decision-making space.

Resource allocation, dispute management and the enforcement of local regulations require sensitivity and special facilitation skills, which forest department personnel lack. Consequently, the department's interventions often disturbed existing arrangements and the enforcement of access restrictions, creating more problems than they solved. One such example from our case studies was Aonlapal Village, where the forester's insensitive intervention led to the breakdown of a well-functioning CFM institution. Here, the village youth club had successfully regenerated the adjoining reserve forest during the 1980s with the passive support of seven adjoining villages and two hamlets. In 1992, a joint committee of the nine villages and hamlets was set up, reflecting the growing community interest in managing a regenerated forest. Tensions developed and tribal villagers felt excluded from the decision-making process. The forest department intervened and registered one of the factions as a *vana samrakhan samiti*, with the promise of generous funds but without understanding the ongoing dynamics. In response, other villagers formed two more *vana samrakhan samitis*, with membership based on political party affiliations. The multi-village collaborative arrangement broke down completely, and some of the villagers were completely excluded.

## Equity concerns in CFM

Traditionally, women have been marginalized in all community affairs and community-level decision-making in Orissa. Their voices have, at best, been represented through their husbands or other male members of the household. CFM systems were no exception to this, despite the fact that the poorest women depended heavily upon forests and forest products.

In all but one of our case study sites, women's involvement in decision-making was marginal. In Aonlapal, Jhargaon and Suruguda villages, women were included only notionally in order to meet the requirements for women's presence in *samitis*, with no real involvement in forest management decisions. The exception was Baghamunda Village. Initially, men used the women as a front to check pressure on forests. Gradually, the women gained confidence and began protesting against men's dominance in decision-making. One remarkable outcome of this process was that the women were included in the local tribal *panchayat* (traditional community forum), although tradition excludes women from such institutions.

In general, CFM has made little difference for poor women since their needs and priorities have not been reflected in local forest management systems. Large farmers, in particular, were more interested in the non-consumptive ecological functions of forests because they gained most from improved soil fertility associated with forest protection. Yet, the costs of protection were borne by those dependent on consumptive uses, both in the form of lost wages for the time spent in patrolling and from restricted access to forest products, especially for fuelwood destined for sale.

The negative impact of elite male control over CFM decision-making on women's access to forest benefits is best illustrated by Gadabanikilo Village in Nayagarh District (Vasundhara, 1997). Prior to CFM, mahua (*Madhuca indica*) seed collectors kept all of the seed they gathered. As mahua seed is primarily collected by women, this was an important source of income and household nutrition for poor women. The male village committee then introduced a rule that 50 per cent of the mahua seed collected from the community-managed forest must be deposited with the committee for equal distribution among all village households. In one sweep, poor women collectors were deprived of 50 per cent of their collection for the benefit of non-collecting households. In 1999, the village committee changed the rule again and auctioned the right to collect mahua seeds to a private individual who offered the highest bid. This individual then employed labour for collecting mahua seed, giving them 50 per cent of the quantity they collected as wages and keeping the other 50 per cent himself. The village committee also restricted the access of NTFP gatherers from other villages, again predominantly women, adversely affecting their livelihoods.

The state JFM order does prescribe representation of different socio-economic sections, including women, in *vana samrakhan samitis*. However, formal representation is inadequate by itself in the absence of explicit recognition of unequal gender and power relations within communities, and firm provisions to ensure that livelihood interests and the rights of the poorest are given priority and protection. In practice, Forest Department field function-aries have often reinforced existing inequalities within communities by strengthening the hands of the rich, rather than siding with the poor. Suruguda Village, in Sundergarh District, provides an example.

Suruguda is a large heterogeneous village that has received wide acclaim and awards for its forest protection efforts. The village started forest protection on its own in 1985; but its efforts were later formalized under JFM around 1990. The *harijans* (lower castes) took the lead in forest protection. As the forest regenerated, the dominant higher-caste majority community of the *agarhias* appropriated control over management decision-making. As a result, forests were 'opened up' for extraction only for a few days, allowing the *agarhia* house-holds – who could hire labour and forego their own income-earning activities in a way that the *harijans* could not – to take full and unfair advantage of extr-action. The Forest Department staff supported the *agarhias*, as reflected in the handling of many forest offence cases. While the *agarhias* got away with cutting even large timber trees, the *harijans* were doubly fined both by the village committee and the forest department for taking even fuelwood.

Such cases highlight the problems with simplistic discussions of 'com-munity' rights that do not ensure the protection of the rights and interests of overlapping groups, such as NTFP gatherers. For many poor women, CFM has only meant a shift in the *danda* (stick), from the hands of the forest guard to the local youth. *Thus, the issue is not just that of who – the forest department or the community – manages the resource, but, more importantly, how resource management is governed within the community.* Unless the management objectives and priorities for community resource management are defined through broad-based participation of forest-dependent women and men in inter- and intra-village forums, mere devolution of existing protection and management systems will have little meaning for poor women. The forest department has done little to address this issue in formulating its JFM policies and has, instead, often reinforced local inequalities.

## Federations of community institutions

As forests regenerate and become more valuable, there is increased pressure on the resource, and conflicts over sharing forest products surface. Com-munities are increasingly faced with situations that require concerted and coordinated responses at larger geographic scales. This has prompted villagers to come together to build alliances.

Villages in Orissa have federated at various levels and in various forms to share experiences, solve problems and gain strength from each other. At a local level, clusters of villages have federated to develop agreements over resource access and sharing among themselves, and to resolve conflicts regarding the trespassing of their respective boundaries. Federations have also been used to enhance the bargaining power of villagers as they deal with external agencies, especially the forest department, over issues of benefit-sharing, management practices and decision-making authority. One important example of collective strength is from Paiksahi Village, where villagers were able to keep a paper manufacturing company out of their forests (see Box 3.3).

---

**Box 3.3** *Strength of collective action: Paiksahi Village, Orissa*

Paikasahi is a small village of 75 households in Ranpur Block, Nayagarh District. The village took up protection of 800 hectares of the degraded Patia Reserve Forest in 1990. In 1996, the forest department formed a *vana samrakhan samiti* in the village and allotted 801.71 hectares of the reserve forest for JFM. JFM enhanced the villagers' sense of control over the forest. However, on 28 October 1997, they found a representative of a paper manufacturing company at their doorsteps. The range forest officer told them that the reserve forest fell under a bamboo working circle and had been leased to Ballarpur Industries Ltd (BILT) for harvesting bamboo. Rights over bamboo from the forest had been granted to the paper company without any prior consultation with the so-called 'joint managers'. JFM's commitment to benefit-sharing with the *samiti* had been totally disregarded.

Paiksahi villagers were determined not to let the paper industry walk away with bamboo from 'their' forests. They decided to protest, supported by other villages in the locality. These villagers formed a Forest and Environment Protection Forum to spearhead the protest. More than 300 villages participated in a joint protest rally on 9 November 1997. They marched silently to the local range office with placards reading '*Aama jungle amara, Purna aain rad karo*' ('Our forest is ours, change old policies'), '*Aame baunsa katai debu nahi*' ('We will not allow the bamboo to be cut'). A memorandum was submitted to the range officer, with copies sent to various senior officials and the chief minister. The forum mobilized support from various quarters, including hundreds of forest-protecting villages from all over Orissa. The case received local and national press coverage. Women from Paikashi and nearby villages participated in the rally, attended meetings and sat with men on roads blocking the entry of industry staff and agents' to the forest. Women leaders from Paikasahi walked from village to village to inform other women in the area about their problem.

Village representatives met the agriculture minister and other political leaders. The minister gave assurances to look into the matter. The response of the principal chief conservator of forests (PCCF) was a major disappointment. He asked the villagers: 'Who asked you to protect the forest? The reserve forests belong to the forest department. The bamboo forests have been given on lease to the paper industry, and villagers can have no rights over these forests.' Following this meeting with the PCCF, on 27 November, some forest department staff visited the village. The villagers were asked to call off the protest and to permit the harvest. They were offered higher wage rates for harvesting the bamboo. When the villagers refused to budge, they were threatened. Two days later, on 29 November, the agents of BILT started harvesting bamboo, using labour hired from distant villages. A tussle between the villagers and the labourers ensued. The tension forced the industry staff to withdraw.

Shaken by the scale of public protest, in December 1997, the divisional forest officer invited local leaders for a discussion. The leaders suggested an open-house discussion and the officer agreed. People insisted on holding this meeting at Kendu

Village instead of at the department's office, and on all the costs for attending the meeting being met. In spite of heavy rain, over 300 village men and women from Paiksahi and other villages gathered. Women who hardly ever attend public meetings shared the floor with village men and officials. They described the pain and effort they had put into protecting their forest and their disappointment over the industry being permitted to harvest the bamboo. They asked: 'We were assured by forest officials that under JFM villagers would have rights over the jointly managed forests. Is *this* joint forest management?'

The villagers also lamented the delay in signing the memorandum of understanding (MoU) between the forest department and villagers for JFM. In the absence of the MoU, the legal position of Paikasahi Village remained weak. The range officer was directed by the divisional forest officer to complete the formalities of registration of the committee within 21 days. The officer assured the villagers that, in future, people would be consulted prior to any management decision such as bamboo harvesting, and their rights would be safeguarded. The villagers felt happy with the meeting's outcome; but this happiness proved short-lived. The 1993 JFM resolution permits 'about 200 hectares' to be allocated to a *vana samrakhan samiti*, while Paiksahi had been protecting 800 hectares of forest. The forest department asked the president of Paiksahi *samiti* to sign a MoU for 200 hectares of the forest. Paiksahi found this unacceptable. Since then, no further action had been taken by the department to formalize the *samiti* or sign a MoU until April 2001.

Paikasahi villagers were successful in preventing bamboo harvesting from Patia Reserve Forest under their protection. Despite their weak legal position, they succeeded due to the strength of collective action. JFM continues to confront hundreds of other villages with such hurdles (Vasundhara, 1998).

These federations represent an advanced stage in the development of community management of natural resources, providing a support system to member CFM groups. They have played an important role in expressing a collective voice for various CFM groups, resolving conflicts and sharing experiences. They have also faced a number of challenges that threaten their continued contribution to creating space for local forest management. Six of these federations were studied in detail by Vasundhara for this research. A brief description of three of these highlights where federations have made progress and what sort of problems they have encountered.

Budhikhamari Joint Forest Protection Committee (BJFPC) is a federation of 95 villages in Mayurbhanj District. The villages primarily came together in 1988 to counter the high pressure from timber smugglers and to provide patrolling support to member villages. Initially, the forest department played an important supportive role. Subsequently, an NGO also came in to provide support. This NGO introduced the system of payments to watch persons and to village youth functioning as cluster-level coordinators. This led to a breakdown of 'voluntary' contributions. With the subsequent withdrawal of NGO support, the federation faced problems in paying for protection and coordination. The federation had problems with the forest department, too, as the department tried to use the federation to promote JFM in the area, which villagers were unwilling to accept. BJFPC is now wary of both NGO and forest department 'support' (see Box 3.4).

---

**Box 3.4** *Formation of the Budhikhamari Joint Forest Protection Committee (BJFPC)*

In 1987, there was an incident of illicit felling of teak trees from Manchabandha Reserve Forest, protected by Budhikhamari and Kailashchandrapur villages. About 25 individuals from these two villages rushed to the forest and stopped the vehicle loaded with smuggled trees. The timber smugglers offered 2000 rupees to the watchers. When this did not work, there was a physical confrontation and 18 people from the forest protecting villages were injured and had to be hospitalized. The smuggler lodged a false case of looting with the police against these 25 individuals and they were arrested. Budhikhamari and the surrounding villages approached the divisional forest officer and the district collector and persuaded them to release the arrested individuals. After this incident, the villagers decided to form a joint committee to deal with timber smugglers. On 30 January 1988, villagers from 20 villages came together to form the BJFPC.

Village volunteers formed a joint mobile patrolling squad. Initially, the squad was 100 persons strong, with 5 persons from each of the 20 villages. By 1998, the number of member villages increased to 95. In 1989, the watchers in two villages faced problems stopping women head loaders. In 1990, women were included in the patrolling squad. Initially, the members of patrolling squad were unpaid volunteers; but from 1992 they were paid. The patrolling squad was very effective in combating timber smuggling in the area. Consequently, in 1992, its strength was reduced to 21 members, including 3 women. Although women were included in the patrolling squad in 1990, they were not included in the executive committee until 1997 to 1998. Unlike men, these women were not in the federation as *representatives* of their villages, or as village leaders, but to demonstrate women's involvement in the federation and to interest other women in forest protection. Until 2000, these women continued to be executive committee members, but as paid watchers having an employee–employer relationship with other committee members, instead of as equals.

---

In Bonai Forest Division, in Sundergarh District, a federation of 128 forest protection groups was formed in 1994. Their experience illustrates the complex relationships CFM federations often develop with state agencies. In this case, the forest protection groups initially came together to try and stall the divisional forest officer's transfer from Bonai, as he had been supportive. The forest department's initial support for the federation waned once the federation started taking up issues that made the administration uncomfortable, such as questioning the granting of leases for stone-quarrying in forest areas and non-enforcement of the rights of *kendu*-leaf pluckers. The withdrawal of department support, coupled with several other internal problems, weakened the federation. It was in the process of revival at the time of this research.

The case from Balangir District illustrates another common problem in the alliance-building process. The process of federation formation has often moved too quickly to give institutional form to existing networking efforts. As a result, the links between these apex bodies and village groups are weak, and democratic mechanisms for representation are not fully evolved. In Balangir District, the Regional Centre for Development Communication (RCDC), an NGO working on natural resources management, was involved in facilitating the formation of a district-level federation. RCDC started the process in 1994. The need for networking was articulated in meetings with villagers. In 1995, a decision was taken by the village representatives and local leaders to form a district forum when the process of alliance-building had commenced in only two of the blocks in the district. The NGO later realized that this step had been taken prematurely. Vasundhara faced similar problems in Ranapur block, Nayagarh District, when a block-level forum in the form of an ad-hoc committee was formed at a meeting, with representatives from only a small number of the total forest-protecting villages in the area. The ad-hoc office bearers of the forum became responsible for initiating processes of internal democracy and representation. Unfortunately, the office bearers then wanted to retain their positions, and in some subtle ways blocked the process of fair representation, distorting the federation's elections in 1999.

The inherent gender and social inequalities of village community institutions were also magnified at higher levels of organization. One rarely finds any women or individuals from weaker sections in organizations representing village clusters. At district and state levels, women or individuals who belong to weaker sections are often completely absent. Thus, the ability of the federations to genuinely represent and articulate the requirements of the real forest-dependent people – women, scheduled castes and scheduled tribes – is severely compromised. The transaction costs involved in being an active leader of a federation, involving meetings at regular intervals and travelling, also make it very difficult for genuine community leaders and forest-dependent poor to take leadership roles. The leadership, in general, passes onto full-time paid persons, teachers, professional politicians and others with less direct dependence on the forest. In the absence of clear mechanisms for downward accountability of the leaders to member CFM groups and village women and men, these leaders cease to be true 'representatives' of their constituent group. A profile of the executive committee members of the two district-level federations – *Jungle Surakshya Mahasangh Nayagarh* and *Zilla Jungle O'Paribesh Surakhya Samiti*, Dhenkanal – revealed the relatively elite backgrounds of committee members.

The weakness of the CFM federations is also evident from the fact that few of them have depended upon internal resources. To some extent this is a reflection of contamination from NGO ideology, as NGOs invariably depend upon donor support for their activities. The process of building up funds through contributions of the people is an important strategy for grassroots mobilization. However, this difficult option has, in general, been given up in view of the easy funding available from NGOs and donors. The NGOs and donors also encourage the same, as funding makes it easier for them to control federations.

The state-level Orissa *jungle manch* (forest forum) is a prime example of institutional form being given to an alliance without adequate groundwork, mainly to cater for the needs of external agencies. The Orissa *jungle manch* was formed at a two-day meeting held in March 1999 in Bhubaneswar. The immed-iate need was to have a body that would deal with the Orissa state government and to represent communities in the sectoral policy and planning unit, to be formed by the forest department, a condition for continued Swedish aid. This *manch* was thus formed before adequate organization at other levels. Ironically, the two spontaneously evolved federations at the sub-district level, the Bonai federation and the BJFPC, were not included as members of the state-level forum. The state-level forum prescribed only district forums as member units. At the time the *manch* was formed, there were only three district forums, all promoted by NGOs and donors.

While these federations provide critical support to their constituents, their acceptance by the state and other external actors as the legitimate voice of CFM offers opportunities, as well as threats. There are clear possibilities of cooption of these emerging institutions by the NGO/donor nexus and the state. In sum, the federations are yet another arena where the struggle for control over forest management is played out.

## Conclusions

Villagers have appropriated significant space for local forest management in Orissa, bringing clear benefits in terms of forest regeneration and the mainten-ance of local livelihoods, although significant problems of social and gender inequalities persist. *Rather than nurturing these local initiatives and facilitating more democratic and gender-equal self-governance by them, the state has imposed a model of JFM that reduces local control over decision-making, endangers local livelihoods of the poor and women, and often hastens the degradation of forests. While JFM does offer formal recognition and some level of technical support valued by CFM groups, the reassertion of forest department control over local initiatives represents an extension of centralization, rather than any devolution of authority and entitlements to local levels.*

The weaknesses of JFM have encouraged local forest management groups to organize and oppose state encroachments on local forest management. Many of these are now formed at larger spatial scales, acting as federations of CFM groups, to tackle the problems ignored or created by JFM. The process of federating, however, brings its own challenges. Co-optation by outsiders and poor representation of the weakest and most forest-dependent sections of the population, particularly women and tribal groups, are the most critical. Here, too, there is room to improve the types of support offered to villagers who have formed their own institutions to manage forests for their own needs.

# MADHYA PRADESH

Madhya Pradesh (MP) is the largest Indian state in area and the sixth largest in population at 66.18 million people in 1991. Forests account for 35 per cent of the state's geographic area and represent 20 per cent of the total forest area of India. Thirty thousand of the state's 71,526 villages are located within or on the fringes of forests. 90 per cent of the state's scheduled tribe population (representing 22 per cent of its total population, and the largest scheduled tribe population among Indian states) live within or near forests. 44 per cent of the state population lives below the poverty line, out of which 80 per cent are concentrated in forest areas (MPFD, undated).

Local people appropriate space for forest management in three forms in Bastar in MP:

1  customary local forest management based on traditional village boundaries;
2  self-initiated CFM facilitated by NGOs and people's social movements; and
3  assertion of the right to self-governance under PESA by a small number of villages, including local control over management of community forest and land resources.

In contrast to Orissa, where the Forest Department presence has been relatively weak, in MP, World Bank funding for a forestry project has substantially strengthened the department's presence and jurisdiction, even in remote forest areas. Although lauded as a success story by the MP Forest Department and the World Bank's evaluation mission, JFM has limited space for local forest management, as explained below.

Unlike Orissa and UP, the MP state government has also initiated an apparently progressive policy. In line with the 73rd Constitutional Amendment, a structure for the decentralization of government to district, block and *panchayat* levels has been put in place. On paper, MP has also devolved more authority to village *gram sabhas* under PESA than other states with Schedule V tribal areas.

Little of this, however, has been translated into practice. Most government schemes continue to be implemented through parallel 'participatory committees', set up in villages by different government departments. These have ambiguous links with Panchayati Raj (local government) institutions and remain answerable to the sector departments that set them up rather than to elected *panchayats* or village assemblies of adult voters. Prominent government-promoted village institutions for forest management include *van suraksha samities* (forest protection committees for well-stocked forests) and *gram van samities* (village forest committees for degraded forests) for JFM; eco-development committees in and around protected areas; *van dhan samities* (in Bastar) for collecting and marketing non-nationalized NTFPs; and primary co-operative societies for collecting and marketing nationalized NTFPs.

## Methods and sites

Case studies were conducted in 13 villages for a more in-depth understanding of the impacts of JFM in MP. In Bastar, where community involvement was already high, the interaction between JFM and existing local management was studied in the Central Bastar and Sukma forest divisions of Jagdalpur Circle. The studied villages were Darbha, Chindawara, Pedawada, Paknar, Chandragiri and Jeeram (in Darbha Range), and Kokawada, Urmapal and Rokel (Tongpal Range) (Sundar, 2000). In addition, four case studies were undertaken in the different context of Harda Forest Division in central MP, where JFM is ten years' old and has been acclaimed for its 'success'. In Harda Forest Division, the functioning of JFM committees in Badwani, Keljhiri and Gorakhal Villages (Rahetgaon range) and Malpon Village (Handia range) were studied (Bhogal and Bhogal, 2000).

## The historical context of Bastar for today's policies

Formerly a princely state, Bastar is sparsely populated, in the *adivasi* (tribal) belt, and has the highest percentage of forest cover (57.25 per cent of its geographic area) in a state known for its forests.[14] Bastar has seen concerted efforts to exploit its forests by the state and private mercantile interests, as well as efforts to preserve its forests by villagers, activists, some elements in the local administration and higher levels of government. Bastar has tropical moist deciduous forests with sal (*Shorea robusta*) dominance in the south, shading into teak (*Tectona grandis*) in the north. A wide diversity of NTFPs provide critical seasonal income and nutrition (in the form of tubers, leaves, fruits and flowers) to the villagers during lean agricultural periods. Forests also provide material for housing, fencing and artisanal production, besides fuelwood, pasture and fodder.

Until independence in 1947, strong traditions of CFM seem to have existed all over Bastar. Unlike recent community initiatives in Orissa, this management rested on the recognition of village boundaries in forests and making offerings to the Earth gods for use of the forest. Villages charged residents of other villages a small fee known as *devsari, saribodi, man* or *dand*, for cutting timber or other produce from their forest.

In 1905, colonial policy initiated reservation of one third of Bastar's total area as forests, while clamping down on shifting cultivation. Reservation involved deportation of some villages, either because they practiced shifting cultivation or because they were simply in good forest coveted by the state. In response to a major rebellion against reservation in 1910, the area to be reserved was reduced to approximately half of that originally planned and the administration was made less intrusive. In villages near the boundaries of the reserves, the state left additional land equal to the existing cultivated area for future extension of cultivation. Twice the existing cultivated area was also left aside for *nistar* purposes. These were forested lands from which villagers could collect forest products for non-commercial household use. At the same time, minor forest produce was redefined as state property and *nistar*, and grazing dues were imposed in 1898. The fight against shifting cultivation continued, however, as did various forms of protest (Sundar, 1997).

In 1949, just two years after independence, all the *nistari* (community) forests of Bastar were also declared government protected forests, with the condition that the *nistari* rights of the people would not be affected. The better blocks were surveyed and demarcated as protected forests under forest department control during the 1960s, while others were left unsurveyed and painted 'orange' in the maps. With control shared uneasily between the revenue and forest departments, the orange areas largely became open access lands. The revenue department continued granting title deeds or *pattas* to land in these orange areas as government policy favoured their distribution to the landless throughout the 1970s. A fresh survey of the orange areas was ordered around 1998 to 1999, and instructions were issued that the areas found unsuitable for demarcation as protected forests could be de-notified and reverted to the revenue department.[15] Despite appropriation by the state, however, villagers have continued to refer to various patches as their *nistari* forests and feel proprietary towards them.

*Post facto* declaration of large areas as state-owned protected forests in the absence of comprehensive surveys and registration of pre-existing cultivation made large numbers of cultivators illegal 'encroachers' on 'forest' land. Whereas earlier, the state government regularized such *de facto* cultivation periodically, the Forest Conservation Act, 1980, (FCA) made prior approval from the central government mandatory for doing so. In Bastar, the issue of encroachments is closely intertwined with the question of basic land rights of the poor. With only about 27.5 per cent of the total area of Bastar (officially) under cultivation and 57.25 per cent declared as state-owned forests, legal access to land for subsistence agriculture has continued to evade the poor, and they form a large share of those labelled as 'illegal' encroachers.

## Surviving traditions of customary forest management in Bastar

Villagers have developed and maintained local institutions for regulating forest use in many villages in Bastar. These institutions have evolved in spite of appropriation of legal control by the state. We describe some of these institutions in our case sites and the types of local decision-making they support.

Ulnar is a large head village of a cluster of 12 villages in Jagdalpur District, which had a *nistari* (mainly sal) forest of approximately 2430 square kilometres that was distributed among the 12 villages. As in much of Bastar, Ulnar had a traditional community-based system for regulating inter-village use of forests.[16] Each village had a forest *sarpanch* and engaged watchmen. If there was excessive felling in any of the villages, the other villages would scold them, saying that if they deforested their own area, they would not be let into any of the other villages' forests.

In 1937, S R Daver, the chief forest officer, formalized this customary multi-village management system of Ulnar's *nistari* forest in a working scheme, which operated until 1952. The scheme divided the forest into seven or eight felling series, each assigned to a set of villages made responsible for its management, including the payment of watchers. Species such as mahua (*Bassia latifolia*), tamarind (*Tamarindus indicus*), hurra (*Terminalia chebula*), mango (*Mangifera indica*) and trees in the sacred grove around the local deity's shrine were not to be cut.

The forest *sarpanches* (heads) collected 1.5 to 2 kilograms per rupee of land revenue, which went towards paying the watchmen, buying uniforms, axes, the construction and repair of the grain depot and meeting other work-related needs.[17] The watchmen were paid 30 to 60 kilograms of paddy per year and were exempted from supplying free labour for landlords or state officials. The forest *sarpanches* met weekly at the market in Bajawand. This system ran successfully for many years as it built upon an existing system. Declaration of *nistari* forests as state-protected forests reduced the efficacy of the system as neighbouring villages stopped recognizing Ulnar's customary authority over these forests. More recently, the forest department had introduced JFM to 2 of the 12 villages in the system without giving any consideration to the surviving multi-village institutional arrangement (see 'Madhya Pradesh's JFM Framework' below).

Junawani Village also had a system of customary management similar to that in Ulnar. Since the 1930s, every household has contributed paddy to employ three watchmen. If they needed additional funds, they sold a tree to a neighbouring village that did not have its own forest. Some of these villages also gave a fee to Junawani for use of its forest. Anyone needing timber for building a house or for a funeral would ask the forest *sarpanch*, who would hold a meeting and generally grant permission. Anybody caught felling without permission was fined. The forest *sarpanches* rotated. In contrast, in some villages around Kanker, the required contribution varied with the amount of timber taken. In others, it was not levied for dry or fallen wood, but only for good timber or only if the wood was stolen. Some villages expected contributions for grazing, while others did not. Apart from Ulnar *pargana* (an administrative unit comprising several villages), other villages in the area also followed a similar system (see Sundar, 2000b, for details).

These customary systems still seem to be fairly effective at the inter-village level, but are clearly under pressure. For instance, somewhat bitterly, Junawani villagers recounted that while the neighbouring villages slowly stopped giving them a fee, they continued to take for their needs from the Junawani forest: 'We don't say anything since people have become educated and tell us that it is not our forest but belongs to the government.' The turning point came during 1983–1984, when Ulnar attempted to prevent illicit felling by villagers from Devda. The dispute was taken to court and, ironically, the police took away Junawani's leaders and a few elders in handcuffs. The case was still going on, but Junawani stopped asking any of the villages for a fee for use of their forest, and protection became lax. By condoning illicit felling by Devda, the police forcefully conveyed to Junawani villagers that their customary authority over their forest no longer had any legitimacy.

## CFM initiatives with outside facilitation

Some villages have developed management systems under the influence of activists or NGOs, systems that often co-exist with customary systems and JFM. Activists of Ekta Parishad, a people's social movement, have organized forest protection in several villages, including Salebhata. In this area, villagers would cut timber from each other's forests and give 2–4 rupees as a fee, following long-established custom. Around 1985 to 1986, under the influence of Parivartan (an NGO affiliated to Ekta Parishad), Salebhata and three neighbouring villages began to protect their own *nistari* forests. Since protection began, the giving of fees ceased because they stopped allowing each other to cut timber at all. Under this system, five or six men from different households patrol the forest in turn. When people need firewood or timber, they apply to the village committee for permission, who inspect the timber after it is cut. There is no ban on grazing. Women also come to meetings and take part in Ekta Parishad rallies.

A consistent feature of CFM in the Bastar case studies is the readiness of women to protect their forests. For instance, in Belgaon, Korkotti, Bade Khauli and some other villages in Keskal range, *mahila mandals* (women's associations) formed by Parivartan began protecting their forests during 1997. In all of these cases, however, the women faced problems from men from their own and neighbouring villages when they attempted to stop large-scale felling by men.

In 1987, the forest department tried to enclose some of the reserve forest adjoining Asna Village near Jagdalpur. This provoked strong resistance by the Asna women, led by Mitkibai and guided by a local activist. Since then, the Asna women had been protecting their forests, and some of them have been successfully running a primary cooperative society to buy NTFPs. This, in turn, led to deep differences within the village and attempts to form an FPC for JFM failed (see Sundar, 1998).

The women of Metawada were assisted by the same activist to form an FPC for JFM in 1994. Initially, they appointed a young man as their president and a woman as the vice-president. For two to three years they all went on night patrols. However, as the president wouldn't call them for the meetings, they decided to have an all-women FPC. As Harawati, the FPC president, said in a village workshop in December 1999: 'You people [men] say what can the women do, but if you don't tell them about the meetings, then really what *can* they do?' Harawati, like other women activists, had faced threats and had even been thrown out of caste for her work.

Bastar clearly had, and continues to have, a wide range of customary and evolving community-based systems for regulating forest use at the village and landscape levels. All have represented the exercise of local authority in institutionalized, widely respected forms.

## Madhya Pradesh's JFM framework

Madhya Pradesh (MP) passed its first JFM order entitled 'Community Participation in Preventing Illicit Felling and Rehabilitation of the Forests' in December 1991. This order was revised in 1995 to coincide with a large World Bank-funded forestry project. The third and latest fourth revisions of the state JFM order were issued on 7 February 2000 and 22 October 2001.[18]

MP has brought the largest forest area under JFM of any state. By the middle of 2000, 6556 village forest committees (VFCs), 5316 FPCs and 323 eco-development committees (for protected areas) had been formed, 'jointly' managing 5.8 million hectares of forest land.[19] This accounted for 37.54 per cent of the state's total forest area of 15.45 million hectares (MPFD, undated). The state's 1995 JFM order stood out among other state orders for making even well-stocked forests eligible for JFM. FPCs within 5 kilometres of well-stocked areas, however, were only entitled to their domestic requirements, with no share of income from timber harvested by the forest department. Village forest committees within 5 kilometres of degraded forests were to get 30 per cent of net income from timber. One man and one woman from every household were to be made general body members of both forest protection and village forest committees.

The February 2000 order went one step further by including even protected areas within its ambit. Although there is to be no benefit-sharing of produce from such areas, according to the order, eco-development committees on the periphery or within protected areas will be entitled to monetary compensation equivalent to the share received by forest protection committees in the vicinity. The proposed benefits to eco-development committees, however, leave unstated the costs imposed on their members in the form of lost forest-based livelihoods. In the new order, members of forest protection and village forest committees will now be entitled to:

- royalty-free *nistar* (forest produce for domestic needs);
- all intermediate products (through cleaning, multiple shoot cutting, etc), both on payment of the forest department's extraction costs (which implies that harvesting of even intermediate products will continue to be undertaken by the forest department rather than the villagers); and
- NTFPs in accordance with government policy framed under PESA.

Forest protection committee members are now also entitled to 10 per cent of the income from final timber/bamboo harvest (compared to none in the 1995 order), while village forest committee members continue to be entitled to 30 per cent. The latest 2001 order has further increased the villagers' entitlements, but has not changed the distribution of power between the villagers and the forest department.

The Madhya Pradesh State Act, passed under the central PESA on 5 December 1997, provides that the *gram sabha* in Schedule V areas shall have the powers to 'manage natural resources including land, water and forests within the area of the village in accordance with its traditions and in harmony with the provisions of the Constitution'. However, this is to be done 'with due regard to the spirit of other relevant laws for the time being in force'. The forest department has used the latter provision to argue that *gram sabha* powers under PESA remain subject to the provisions of existing forestry legislation.

MP's February 2000 and October 2001 JFM orders incorporated PESA provisions in making all adult voters of the *gram sabha* eligible for general body membership of the three types of committee (in contrast to one man and one woman per household in the 1995 order). Another significant provision is that committee members will be treated as public servants while on patrolling duty and are entitled to legal protection and the same compensation as forest staff in case of death or injury.

The World Bank-funded Madhya Pradesh Forestry Project provided alternative development inputs to villagers to wean them away from the forests through 'eco-development', in the case of protected areas, and the Village Resource Development Programme in JFM villages.[20, 21] The development programme funds allocated to each participating village were 300 rupees per hectare per year over seven years for a maximum area of 300 hectares per village. Up to one third of its annual entitlement was to be paid to each village forest committee on a monthly basis, either directly or into its bank account. In theory, the committee could use this money either to pay wages for protection or for community benefit, provided it took care of protection through voluntary effort. 700 million rupees were credited to committee bank accounts by the middle of 2000 (MPFD, undated). The remaining two-thirds of the village forest committee's annual entitlement was to be used for village development works, such as minor irrigation, for which the divisional forest officer made direct payment, although decentralization to local self-governments requires that these responsibilities are assigned to Panchayati Raj institutions.

## Impacts of JFM in Bastar and Harda

### *Organizational inclusiveness*

All nine JFM villages studied in Bastar failed to meet the membership norms prescribed by MP's 1995 JFM order of including one man and one woman from every household. According to the data supplied by the divisional forest officer, for all 24 JFM groups formed in Darbha Range, the number of women members varied from only 2 to 10, while the number of men varied from 6 to 110. One of the deputy rangers said that although they had attempted to involve the entire village, membership wasn't such an issue yet since distribution of the harvest had not begun. JFM committee meetings were held on a random basis every three to five months, although one of the divisional forest officers had recently ordered that they should be held once a month. In Harda Forest Division, however, one man and one woman from each household had at least been listed as members – although the women, in all four cases, knew little about their committees.

The problem was not just a failure to include women and other disempowered groups in the forest protection and village forest committees. Superimposition of JFM committees on pre-existing CFM systems in Bastar had, in some cases, made decision-making much less inclusive. For example, in Belgaon, the *mahila mandal* was protecting a forest in cooperation with two other villages. At the initial meeting for forest protection committee formation, all households were invited and the signatures of those who came were taken. However, they were told nothing about JFM rules and the division of roles and responsibilities between the committee and the forest department. The committee had received money to trade in tamarind and urea, doubling up as a *van dhan samiti* (NTFP marketing committee); but no one, except the office bearers, knew anything either about the trading or the accounts. The women's major complaint, however, was that the committee president took money on behalf of the committee for allowing people from other villages to cut trees from their forest, kept it for himself and didn't inform them. When they tried to stop offenders, they were told that money had already been paid and that they could do nothing. In this case, JFM reduced rather than increased forest-dependent women's control over their forest.

## Democratic village leadership or agents of the forest department?

In contrast to the official claim that villagers have been substantially empowered to take forest management decisions through the formation of village organizations under JFM, our research found that the majority of villagers in all JFM sites had little knowledge of the decision-making process within the committees. Where attempts were made by villagers to participate actively in forest protection committee decisions, the forest department often thwarted the effort by controlling committee leadership appointments, record-keeping and other key decisions.

This was the case even in the celebrated Harda Forest Division, which piloted JFM in MP during the early 1990s. While the divisional forest officer who pioneered the 'Harda' model may have been personally committed to villagers' empowerment, his successors reverted to old patterns of centralized control embedded in the forest department's unchanged structure. As the secretaries and joint account holders of forest protection committees, beat guards decided which decisions and accounts were recorded and how 'community funds' were spent. They even controlled the appointment of committee presidents. Members of the JFM committee of Gorakhal Village, for example, elected a new president in February 2000 to replace the one who had held that post since the committee's constitution in 1992. Within three weeks, however, the beat guard reinstated the earlier president as he found the villagers' representative too independent for his liking. The residents of Gorakhal Village were visibly annoyed with their forest protection committee. Their committee had approximately 117,000 rupees of its own funds; but requests for loans for bullocks or healthcare were turned down. The villagers claimed not to know who the president of their forest protection committee was; but a man close to the beat guard asserted that he was the president (Bhogal and Bhogal, 2000). Even after ten years, JFM in Harda had failed to develop broad-based democratic village institutions capable of managing forest resources on a democratic, sustainable and equitable basis.

Efforts to maintain the forest department's control are equally evident from the manner in which some potentially empowering provisions of PESA have been incorporated within the state's February 2000 and 2001 JFM orders. The essence of PESA is to promote democratic *gram sabhas* (village assemblies) that make their *own* collective decisions, implying that *gram sabhas* define the boundaries of their community (including forest) resources and take over their management, as is performed by some villages in Nagari Block. The JFM order, on the contrary, empowers the forest department to constitute *gram sabhas* as JFM committees, with the guard or forester as member secretary/joint account holder. The JFM order also vests the authority to allocate forests to JFM committees and to approve and supervise village management plans (including non-forestry development works) with the divisional forest officer. In the case of well-stocked forests, the forest department retains the right over 90 per cent of the income from timber, even from *nistari* (community) forests within revenue village boundaries, and to unilaterally dissolve committees, in which case members lose all of their entitlements to the promised benefits. In effect, this implies that the department can dissolve the constitutionally empowered *gram sabhas* themselves. *The legal empowerment of* gram sabhas *for democratic decision-making by PESA is thus potentially subverted by the forest department's administrative order.*

## Control over decision-making related to 'community' funds

Of all areas in which JFM has constrained the space for local decision-making, control over community funds is perhaps the most egregious. The vast majority of villagers in our case sites knew nothing about the total budgetary allocation for their villages, or that they had the freedom to decide how to use their own share of funds. The forest department, when challenged, claimed the authority to make financial decisions on behalf of villagers on the grounds of being accountable for their 'judicious' use.

The situation was particularly bad in Harda Forest Division. In February 2000, the divisional forest officer issued instructions that no JFM committee could use its own funds without his approval. If the officer did not consider the proposed expenditure to be 'wise', approval was refused. Rotation of committee funds had been minimal, and disproportionate benefits of loans had been taken by committee presidents. Committee accounts held jointly by the committee president and the beat guard facilitate such collusion. At a public hearing organized by *Shramik Adivasi Sanghatan,* a local mass tribal organization, women and men of 12 different villages complained bitterly about not being provided any accounts of their common funds by the forest guard and being denied access to them even for emergency loans. In response to protest rallies, the conservator of forests promised to give them detailed accounts in April 2001; but not a single JFM committee in the area had been provided with any accounts until March 2002.[20] A major demand of the villagers at the public hearing was that their JFM committees should be disbanded, as they had become tools for increasing oppression and exploitation of poor *adivasis* of the area.

Similar mechanisms for maintaining Forest Department control over 'community' funds were evident from the case studies in Bastar. In Darbha Village, for example, the executive committee members were all petty politicians and shopkeepers, generally appointed by the Forest Department. Such people were close to the forest staff and could be expected to collude with them on fund management through the jointly held accounts. One of the major complaints of the Communist Party of India about JFM was that the Forest Department purposely chose non-literate youth as committee presidents when a sufficient number of educated youth were available. In Kandanar, for instance, the forest protection committee president was non-literate, and the registers and passbook were kept with the beat guard. Kandanar's president asked to be told what he was signing and how much money was being withdrawn from the committee account; but the department staff refused. The foresters claimed that they were responsible for ensuring that 'government' funds were not misused. In Chirwada and Badanpal in Tongpal range, the presidents signed away cheques for 25,000 rupees, but did not know the name of the forester to whom they had entrusted the funds or their bank account numbers. The money was to be used for a microphone set and vessels for the village; but the deputy ranger bought these on his own (and refused to tell the villagers the cost). The presidents were upset that they had not been taken along. The forest staff claimed that they had been called but had not come, and that even when they did come, they were expected to 'be treated like *baraatis* [pampered members of the bridegroom's marriage party]' and have their travel paid for.

Even the decision of whether to pay villagers for protection or not tended to be controlled by the Forest Department. In Badanpal, Tongpal Range, four watchmen were not paid for three months' work. The villagers then started rotational patrolling on their own. When asked, the Forest Department staff said that the watchmen were not paid because they did no work and money had to be saved for the committee. The committee members had not even been asked if they wanted to save money in this way.

The effect in many areas was to eliminate the incentives for forest protection that helped to support the local management systems developed by the villagers themselves. Where fines and fees were once collected by local institutions to support patrols and regulate forest use, the Forest Department now determines how even the villagers' money will be spent – often without even minimal transparency. JFM has undermined local incentives to protect forests and has wrested decision-making control over financial matters from local people.

## Negative livelihood impacts and equity

Discussions on JFM largely centre on the package of 'incentives' offered to villagers, such as shares of timber and NTFP revenues, in order to assist the Forest Department in protecting forests. What is less often discussed are the constraints imposed by JFM on the pursuit of pre-existing, forest-based livelihoods. In most instances of our fieldwork, JFM imposed – or tried to impose and failed – restrictions on the activities crucial to the survival of the poor.

Under the Forest Department's influence, for example, most JFM case study villages in Bastar tried to ban sales of bamboo shoots. In Kukanar, however, the specter of the ban was one of the major obstacles to JFM, since poor women depended upon selling bamboo shoots for household food security. Similarly, in Chandragiri, the forest protection committee members said that when they started, protection had been quite strong but declined subsequently, since 'we don't like catching people from our own village' who clearly need access to forest resources for survival. In Darbha and Chandragiri, JFM committees allowed their own villagers to extract timber for building houses and implements, while the Forest Department turned a blind eye. This reflected the villagers' ability to retain at least some decision-making space within the JFM framework, especially where they were better organized. In contrast, in Harda there was a concerted campaign against both firewood headloading and grazing in forests during the early years of JFM. In one of the most well-known village forest committees, in Badwani, a rigorous ban on keeping goats had been enforced since 1991. Poorer village women resented this, saying that goats were like liquid cash and they could manage them so as not to harm the forest. However, they had no say in the matter (Bhogal and Bhogal, 2000). In general, there was no discussion of the inequitable distribution of the livelihood *costs* of 'joint' forest management on those dependent upon daily extraction. Consequently, there was no discussion of more livelihood-sensitive management alternatives.

At the same time, when benefits have been distributed, they have not been distributed equitably. During the May 2001 public hearing in Harda, the assembled villagers accused the Forest Department of co-opting a few better-off villagers to deprive the rest of forest access for subsistence uses. The inequitable distribution of benefits from Village Resource Development Programme investments within villages is evident from the Forest Department's own accounting: 40.8 million rupees had been spent by the department in 145 villages in Harda Division over ten years in order to provide 818 households with irrigation (Dubey, 2001). This amounts to an average of five to six households per village receiving a benefit of almost 50,000 rupees each, when the average number of households in each village is about 100. Ensuring equitable distribution of investment benefits among the majority of village households has clearly not been a priority. While JFM forests regenerate by excluding the poor, the department itself continues 'scientific' fellings for state revenue. In an interesting caricature of state–community relations under JFM, mass tribal organizations have depicted JFM as a cow fed by the villagers but milked by the Forest Department.

JFM had also threatened livelihoods by cutting off forest-poor communities from neighbouring forests, exacerbating inter-village inequality in the process. In all of the JFM villages studied in Bastar, poor villagers from neighbouring villages had their tools confiscated while collecting bamboo, even when they enjoyed long-standing *nistar* rights in the forests brought under JFM. Although this had not led to any major fights, mass tribal organizations allege that conflict between villages has increased.

## Employment and other benefits

One of the early benefits of JFM in villages funded by the forestry project was wage employment. The Forest Department in Harda facilitated villagers' access to a wide range of development schemes within other departments, such as loans/subsidies for house construction, wells and lift irrigation, even prior to the forestry project. All four case study villages in Harda division benefited substantially from improved agricultural productivity and village infrastructure development.

In general, however, already well-off JFM committee presidents cornered the maximum benefits in return for cooperating with Forest Department staff in preventing forest use, often of the kind most crucial to the poorest villagers. Such funding has also not nurtured villagers' informed participation in forest management decisions. In the words of the president of Malpon's FPC: 'The committee will get finished the day the government stops giving it money' (Bhogal and Bhogal, 2000).

Inter-village inequities were also exacerbated. Villagers in Bastar received daily wages of 60–70 rupees for assisted natural regeneration work, but only in villages covered by the Madhya Pradesh Forestry Project. Neighbouring non-project villages received only the money that they collected through fines or, at best, a monthly salary for one watchperson, which could be used flexibly. In Salebhata and Mandri, for example, JFM committees were superimposed on the villagers' pre-existing NGO-facilitated protection committees. Salebhata was allocated a patch of a protected forest that villagers were already protecting. While Salebhata received no money from the Forest Department, the village forest committee of Mandri Village received funds to build a check dam, well and pond. Mandri also received money for a plantation in the reserve forest from the World Bank-funded project. This led to a lot of confusion and suspicion of the Forest Department. Village selection for project benefits seemed arbitrary, and villages that did not receive any funds suspected the department of depriving them of their due entitlements.

Project funds created other problems, as well. Where pre-existing voluntary protection was replaced with paid watchmen, the department's money reduced psychological controls. Some villagers felt that since the department was now paying for protection, they could cut the forests. As long as the villagers were contributing themselves, the psychological restraints were stronger.

*The Forest Department also viewed the JFM committees as instruments for implementing other government schemes, rather than democratic self-governing institutions making their own decisions.* In Darbha, for example, the Forest Department empowered JFM committees as *van dhan samitis* to trade in NTFPs without informing the larger community. The JFM committee money that belonged to the whole village, including funds coming in as the forest protection committee's share of JFM income, was being used for the personal gain of executive committee members as members of the *van dhan samiti.*

## Consolidation of Forest Department control over nistari forests through JFM

The loss of local decision-making space and livelihood options was most obvious in the use of JFM to extend Forest Department control over undemarcated ex-*nistari* forests. Until 1949, villagers had enjoyed formally recognized rights and the authority to use these forests to meet their basic needs, with little interference from the state. Despite their declaration as government-protected forests, the villagers continued to feel proprietary towards them due to lack of surveys and demarcation on the ground. Once brought under JFM, however, local use became subject to the Forest Department's control.

Thus, in Tongpal Range, 5 of the 16 JFM committees were for land formerly used for *nistar*, and were likely to be converted to demarcated protected forests under Forest Department control. In the process, other land uses that had developed over the years were labelled 'encroachments' and, in some cases, removed and replaced with plantations. In Junawani, the department established a 50-hectare plantation on village revenue land where villagers had been growing pulses and oil seeds to supplement paddy. In Kukanar, demarcation of two orange areas had been completed. The village was divided on the benefits of JFM. One view was that JFM would involve the department in taking away villagers' land. Another view compared it to sharecropping – the village owned the land, but since the department would put in money (through payment of wages), it was entitled to 70 per cent of the harvest. *The latter view was obviously uninformed about the likely conversion of their nistari forest into a demarcated protected forest following plantation, making the department the undisputed owner of the land. Dissemination of such critical information to the villagers was conspicuous by its absence.*

The formation of JFM committees by the Forest Department in Junawani and Ulnar also brought the villagers' customary, multi-village *nistari* forest management under the ambit of JFM. *Instead of being the managers of their own community forests, villagers were reduced to 'beneficiaries' participating in the management of a state forest.* At the same time, because JFM was introduced in only 2 of the customary cluster of 12 villages around Ulnar, the department undermined the inter-village access and regulation system for a larger resource landscape. This generated conflict and confusion among the villages over customary versus Forest Department authority.

## Local appropriation of space in the face of JFM

The Forest Department has not had it all its own way in introducing JFM. Villagers have often fought off its implementation or have adapted its provisions to suit local needs. In many cases, there was an adaptive co-existence of traditional authority with that of the Forest Department. We provide a few salient examples among the many that reflect the continued creativity and assertiveness of local people.

In Junawani, the official structure and actual functioning of the village forest committee comprised a mix of traditional and official norms. According to villagers, the committee officially had about 10 female and 10 male members; but according to the minutes of the register, about 40 or 50 people actively participated in each meeting. The Forest Department staff did not follow the JFM order's membership prescriptions, designed to ensure broad-based participation. However, villagers participated anyway due to the strong tradition of collective decision-making by the village assembly. In Ulnar, there was conflict over who should be the protection committee's chair. The Forest Department's nominee was replaced by one camp of villagers, who felt that he was being too strict and wanted someone of their own. The department's appointee continued to sign registers; but the bulk of the work of the chair was done by the other man. The customary system for satisfying the villagers' timber needs continued to function despite having no recognition within JFM. Villagers were charged a fee of 2000–3000 rupees depending upon what they cut. In 1998, eight houses were built and 16,000 rupees were collected.

Although most villagers were not aware of PESA's provisions, in the villages mobilized by Bharat Jan Andolan (a people's social movement) in Nagari block, Raipur District, villagers had begun asserting community control over their forests. In Chanagaon, during 'earlier times', if anyone wanted wood they would ask the headman. After the *nistari* forests were declared protected forests in 1949, the feeling of ownership ceased and everyone cut freely. The Forest Department also cut coupes. Once the village was mobilized by Bharat Jan Andolan, villagers decided to take only as much wood as they needed. The department wanted to form a JFM committee in their village; but the villagers rejected it after studying the JFM order. The 30 per cent share of revenues from timber was considered a poor deal when they could keep 100 per cent of it under PESA. In addition, JFM had no provision for timber for villagers' own use, and villagers asked: 'Who doesn't need to build a house?' They decided it was better to have their earlier system under which people asked for timber when they needed it.

In Uraiya, another Bharat Jan Andolan village, when the *patwari* (lowest revenue official) and forest guard attempted to charge a youth who cut a tree in his own land with violating a tree-felling ban, the villagers took recourse to PESA and approved the felling. Villagers in Uraiya have also taken over patrolling of their erstwhile *nistari* forest from the beat guard, and have been sanctioning wood-cutting on application to the *gram sabha*.

## Conclusion

Most government schemes continue to be implemented through parallel 'participatory committees' set up in villages by different government departments. These have ambiguous links, if any, with Panchayati Raj (local government) institutions and remain answerable to the sector departments that set them up, rather than to elected *panchayats* or *gram sabhas*. JFM in MP, and particularly in Bastar, was almost entirely a product of a top-down process initiated by the World Bank and the state government. The JFM committees were tightly controlled by government staff and rarely represented forest users. In setting up committees for forest protection, there has been no attempt to build upon or formalize existing methods and institutions. Indeed, by imposing a much more restrictive and standard formal JFM structure, the government has further taken away initiative from the villagers. Devolution has become another means to extend Forest Department control to new areas, including the *nistari* forests that villagers consider theirs.

Our studies indicate, firstly, that JFM has undermined communal property rights regimes defined by customary village boundaries, thereby generating inter-village conflicts and inequities. Secondly, it has consolidated Forest Department jurisdiction over *nistari* forests through their demarcation as protected forests. Thirdly, it has destroyed survival livelihoods of impoverished 'encroachers' by converting *de facto* subsistence agricultural land use into forestry plantations. Fourthly, it has shifted the locus of control over decision-making from within the community to Forest Department functionaries, despite the department's historical lack of accountability and transparency. Instead of enhancing multidimensional space for local forest management, these interventions empower the Forest Department, often in alliance with male village elites, to reassert its control over local land and forest resources.

It can be claimed that forest protection has improved in some areas since the formation of JFM committees, and that the Forest Department has acted as a third party when offenders challenged the authority of village guards. Yet, it is in these same areas that the department exacerbated inter-village relations in the first place, weakening existing systems for managing forests at a multi-village landscape level by centralizing forest management and converting *nistari* to protected forests.

In theory, PESA empowers *gram sabhas* to manage community resources, and represents a new and important alternative to the 'committee' approach of JFM. In practice, however, there is little evidence of state commitment to operationalize PESA's empowering provisions. MP's February 2000 and October 2001 JFM orders subsume *gram sabha* authority under Forest Department control. Rather than representing community control mandated by PESA, these policies, at best, represent welfare interventions by the state within new structures of bureaucratic management and control.

People's social movements, including Ekta Parishad, have condemned the Madhya Pradesh Forestry Project and its support of JFM and eco-development. They see such projects as an underhanded means of displacing villagers from *nistari* forests and agricultural lands, encouraging the formation of more plantations that benefit industries and procurring more jeeps and equipment for the Forest Department staff. Ultimately, they argue, JFM, at least in the current version closely identified with the World Bank, is a reformist illusion aimed at diverting attention from the demand for effective people's control. According to P V Rajgopal, the leader of Ekta Parishad, the government must first regularize land under cultivation by tribal and other poor communities, and then initiate a public debate on how much land should be reserved for forests. Ekta Parishad is not against forest conservation. Several of the villages where it is active in Bastar have been protecting their forests much before JFM was introduced. The underlying issue is of rationalizing land use based on balancing conservation and livelihood needs. In other words, meaningful devolution of forest management cannot take place without resolving the basic issue in the lives of most poor Indian villagers: access to land as a means of livelihood.

While PESA has the potential to bring about a radical shift in state–people relations, its introduction has been troubled. The real force behind this act was a handful of people who wanted to enable tribal communities to govern themselves. The tribal communities themselves were the least aware of it. A better-worded act – emanating from, rather than preceding, public mobilization – might have been more effective.

## UTTARAKHAND: FROM RIGHTS HOLDERS TO 'BENEFICIARIES'?

Uttarakhand, the hill region of UP, provides numerous examples of officially constituted and informal CFM systems. According to the 1991 census, Uttarakhand's population was 5.93 million, 78 per cent of whom lived in rural areas. Whereas only 12.6 per cent of the hill region's area is officially under cultivation (Saxena, 1995b), the rural population actually uses about 60 per cent of the total area for sustaining local agro-pastoral livelihoods. Most of this uncultivated land is legally classified as state-owned 'forests', about 67 per cent of the total area of Uttarakhand (Ghildyal and Banerjee, 1998). Between 16 per cent and 20 per cent of the region has been brought under the protected area network under the WPA. The area is about four times the percentage at the national level.

Village communities are more homogenous, compared to high social stratification in the plains. The percentage of scheduled caste population, as well as caste-based exclusion, is lower than in the plains. Land distribution is relatively equal, with rare cases of land holdings of over 2 hectares, and landlessness is low. The area's agro-pastoral economy is still predominantly subsistence-based, with about 50 per cent of rural households, including the rural elite, having high dependence on village commons and forest lands. Whereas the area had a self-sufficient economy at the time of colonial occupation (Guha, 1989), today around 45 per cent of the economically productive work force works outside of the region due to lack of local employment. High male out-migration in search of employment leaves women as effective managers of the rural household economy. About 40 per cent of households are headed by women (CECI, 1998, quoted in Ecotech, 1999).

Uttarakhand consists of two sub-regions: Kumaon and Garhwal. With the advent of colonial rule, Kumaon and a part of Garhwal (referred to as British Garhwal) were brought under British rule, whereas Tehri Garhwal remained a princely state with British support. Appropriation of the uncultivated commons under the two administrations followed different trajectories, resulting in the two having somewhat different laws and land classifications. The rules for *van panchayats,* for example, were framed by the British and were initially applicable only to the territory under British rule. They were extended to Tehri Garhwal only in 1991. The historical analysis here focuses on colonial interventions in Kumaon and British Garhwal, referred to only as 'Kumaon' for brevity, as developments in Tehri Garhwal often followed those in the British-ruled territory.

Democratic and autonomous community management of legally demarcated village forests by elected forest councils, or *van panchayats,* has existed in Uttarakhand for over seven decades. Unofficial community management, with diverse institutional arrangements on all legal categories of forest lands, has co-existed with formally constituted *van panchayats* and, in fact, predates them.

Both the *van panchayats* and unofficial CFM systems, however, confront challenges that result from the new systems being imposed by the state. With funding from the World Bank, the Uttar Pradesh Forest Department has promoted village forest joint management (VFJM) with *van panchayats* on land not under Forest Department control. This is in contrast to JFM on degraded reserve and protected forests under Forest Department *jurisdiction* in other states. This policy encourages the opposite of devolution, creating space for the Forest Department to intrude on village forests managed autonomously by communities, instead of creating space for villagers to participate in managing reserve forests under departmental jurisdiction. *It is misleading to refer to UP's VFJM approach as JFM as it hides this crucial difference from JFM in other states – that* van panchayats *had enjoyed autonomous authority over village forests prior to VFJM.* The decision-making autonomy of *van panchayats* participating in VFJM is 'subject to the supervision, direction, control and concurrence of the divisional forest officer' (FDUP, 1997, p3.1). A functionary of the Forest Department has been made the joint account holder and the member secretary of *van panchayats* (GoU 2001), after having no role for 70 years.

At the same time, informal community management is under pressure from state-driven *van panchayat* formation. The revenue department is demarcating civil forest lands (falling under the legal category of unclassed forests) under its jurisdiction as village forests to be managed by officially constituted *van panchayats.* The department is also dividing existing multi-village *van panchayats* into single village ones, often generating inter-village conflicts and inequity. Eco-development committees, promoted by the wildlife wing of the

Forest Department, solicit villagers' 'participation' in replacing their existing forest-based livelihoods with untested, non-forest dependent alternatives. Decisions taken earlier by villagers – such as whether to take up community management, whether to do so officially or unofficially, and whether to do so at a hamlet, village or multiple-village level – are now being taken by the state on their behalf through target-driven formation of *van panchayats*, eco-development committees and village forest committees for VFJM.

This confusing, often contradictory, array of devolution policies is meeting with responses of forest-dependent women and men that vary from outright rejection to (often opportunistic) acceptance. Unlike Orissa, however, inter-village federations were either absent or very weak. NGOs and civil society groups, which earlier played an important role in policy advocacy, now largely work as 'private service providers' for the many donor-funded projects in the region.

## Sites

Our 16 cases were spread over 9 out of 12 districts of Uttarakhand. We selected the case studies to include self-initiated CFM; old and young *van panchayats* not yet brought under any of the recent devolution policies, both with and without active women's participation; villages that have rejected VFJM; VFJM villages considered 'good' by the Forest Department and those facing problems; villages where new *van panchayats* have been formed or existing ones recently subdivided; and villages targeted by multiple devolution policies.

## Historical context

Prior to British conquest in 1815, the hill peasantry effectively exercised direct control over the use and management of cultivated lands and uncultivated commons, with little interference from earlier rulers. Resident communities regulated use within customary village boundaries by evolving rules rooted in cultural traditions (Guha, 1989; Somanathan, 1991; Agarwal, 1996). Hill agriculture and animal husbandry depended upon spatially and temporally integrated use of cultivated and uncultivated lands. Seasonal transhumance to alpine pastures and grasslands allowed for resource recovery, and high dependence upon forests generated conservation practices that were integrated within cultural and religious traditions, including the maintenance of sacred groves (Guha, 1989).

Interventions during colonial rule permanently altered this landscape of holistic resource management. In 1823, the colonial regime undertook the first land revenue settlement in Kumaon and British Garhwal. This settlement recorded customary village boundaries, categorizing the land within them as cultivated *naap* (measured) and uncultivated *benaap* (unmeasured) lands. Although villagers continued to enjoy unrestricted use and the right to clear new unmeasured land for cultivation, the state appropriated the authority to grant recognition to village boundaries from local institutions. These 1823 boundaries continue to be the basis of inter-village boundary disputes over rights in the commons, including in forest areas reserved 90 years ago, despite several boundary changes that were subsequently introduced (Nanda, 1999).

In 1893, all unmeasured 'wastelands' were declared *district protected forests* (the current category of protected forests). This legally classified all village common lands as 'forests', irrespective of whether they had tree cover or not, and converted them into state property. A resource base that was managed holistically was artificially divided into *forest* and *non-forest* lands.

From 1910 to 1917, the colonial government attempted to tighten its control over forest resources by designating over 7500 square kilometres of the commons as reserve forests, severely restricting people's use rights. Following rebellions and incendiarism,[22] 4460 square kilometres of the commercially less valuable new reserves, classified as Class I Reserve Forests, were transferred back to the civil administration from the Forest Department, and people's rights to them were restored. However, rights in these Class I reserves were given to 'all bonafide residents of Kumaon', thereby converting common property resources defined by the 1823 village boundaries into open-access areas. Provision for *van panchayats* to exercise community control over legally constituted 'village forests', demarcated from within the Class I reserves, and civil forests (comprising the residual area of the district protected forests after reservation) was made, though only in those villages that applied for them. This enabled sections of the peasantry to retrieve some space for local forest management. The state, however, consolidated its control over the commercially valuable forests, classified as Class II Reserve Forests.

Thus, by the early 20th century, the uncultivated commons had been divided into three legal categories of forests: commercially valuable Class II reserves under the Forest Department; commercially less valuable Class I reserves under the civil administration; and civil (*soyam*)[23] protected forests, also under the civil administration. A fourth category of *panchayati* forests (now 'village' forests under section 28 of the IFA could be carved out of the Class I reserves and civil forests to be brought under community management. This did not appease villagers. The state commenced large-scale commercial fellings, and protests continued.

After independence, the state continued commercial forest exploitation with even greater vigour. The reach of the Forest Department and its contractors spread to the remotest corners with the expansion of the road network. Local livelihoods received even less attention than under colonial rule. State policy consistently favoured the export of raw timber and resin for processing by large industry in the plains. By the 1970s, the *Chipko* Movement had emerged to demand that priority be given to local employment in the extraction and processing of forest produce (Guha, 1989). Increasing incidents of landslides and floods, and declining availability of biomass for subsistence needs, propelled even hill women into the movement, broadening the popular base of the *Chipko* protests and giving them their 'eco-feminist' label.

The issue of local forest rights, however, was soon subsumed within the new national and global ideology of environmental conservation. Instead of priority to local forest-based livelihoods and employment, *Chipko* was used to justify a spate of centralizing environmental policies and laws. The FCA empowered the central government to control the alienation of even the smallest patch of forest land. In 1981, central government imposed a 15-year ban (since extended) on all commercial felling in the UP Himalayas above 1000 metres.[24] Today, the only permitted fellings are for the villagers' timber rights, recorded under the forest settlements of 1910 to 1917. These settlements specified village allocations of timber and are now completely outdated and inadequate. Rapid expansion of the protected area network has resulted in additional large-scale resource displacement, affecting the livelihoods of an estimated half a million people. Resal, one of our case study villages located inside the Binsar Wildlife Sanctuary, is left with half of its original population through involuntary resettlement due to wildlife damage to crops and live-stock and constant threats to human life. Women who earned international fame for stopping contractors from felling their forests during *Chipko* have come to hate the word 'environment'. As one of these women from Reni Village complained: 'They have put this entire (surrounding forest) area under the Nanda Devi Biosphere Reserve. I can't even pick herbs to treat a stomach ache any more' (Mitra, 1993).

Over time, the Forest Department consolidated its control over the original forest classifications, blurring the distinctions between them. Today, all state-owned forest lands fall under one of the three categories specified in the IFA – reserve, protected or village forests. The distinction between Class I and II reserves became redundant after most of the former were transferred back to the Forest Department in 1964. All the civil/*soyam* forests (originally called district protected forests) are now equated with undemarcated protected forests under the IFA.

Table 3.2 indicates the area currently under different legal categories of forests in Uttarakhand.

Centralized forest management based on a conservationist ideology further intensified community alienation in Uttarakhand, which significantly encouraged the movement for a separate hill state. It remains to be seen, however, whether the hill peasantry will gain a greater voice in defining the new state of Uttaranachal's land-use and forest management policies.

## Informal Community Forest Management

Uttarakhand has widespread CFM outside of any formal legal framework on all categories of forest lands. Traditional *lath* (stick) *panchayats*, informal forest committees and, more recently, increasing numbers of *mahila mangal dals* (village women's associations) are regenerating and regulating the use of reserve and civil/*soyam* forest lands, often compelling unofficial cooperation by forest and revenue department staff. The majority of these systems rest upon the region's strong traditions of communal resource management based

Table 3.2 *Area of different legal categories of forests in Uttarakhand*

| Category | Jurisdiction | Area (hectares) | Percentage of total forest area[25] |
|---|---|---|---|
| Reserved forest (merged Class I and II) | Forest department | 2,375,571 | 68.92 |
| Civil and *soyam* forests (equivalent of undemarcated or unclassed protected forests) | Revenue department with *Gram Panchayats* | 578,550 | 16.78 |
| *(Van) panchayat* forests (equivalent of village forests) | Van panchayats (with revenue department; in a few cases, with Forest Department) | 469,326 | 13.63 |
| Private, cantonment and others | Miscellaneous | 23,262 | 0.67 |
| Total | | 3,446,655 | 100.00 |

*Source:* Jena et al, 1997

upon customary boundaries that have survived despite more than one and a half centuries of state interventions.

Holta, one of our case study villages, initiated protection of its *soyam* land in 1986 entirely on its own. Village water sources had dried up and firewood and fodder had become scarce as a result of unregulated forest use by surrounding villages and encroachment on communal land. Village youth successfully persuaded the encroachers to vacate the commons, setting an example by giving up their own encroachments. Letters were sent to the *gram panchayat* heads of surrounding villages, stating that anyone entering the forest would be fined. Major conflicts followed, with one village going to court against Holta over unclear boundaries of their respective *soyam* lands.

However, with improvement in forest condition and availability of water, resistance declined. At the time of our research, all of the village's biomass needs, except timber, were met from the regenerated forest. Vegetable cultivation had become feasible with regeneration of three natural water sources. Rules had been framed for grass, tree-leaf fodder and firewood collection and were strictly enforced, with all households contributing to pay a watchman. The committee had representatives from all hamlets and castes, and representatives of the village women's association had also wedged their way in. Community relations with the Forest Department, however, were extremely sour. In the words of the village women, the department had made them into thieves (Gairola, 1999a).

In at least 5 of our other 15 case studies, informal community management was in place. In Makku and Bareth, women's groups had asserted informal control over patches of civil or communal land nearer their settlements. In both cases, the women perceived local *van panchayat* councils to be male-dominated.

*Panchayat* forests were also far from the villages, and were therefore not convenient for daily fuelwood and fodder collection. The formal and informal CFM arrangements complemented each other, with the women occupying an informal space. They secured access with mediation of the *gram panchayat* without having to deal with cumbersome official procedures. In Arakot Village, the *mahila mangal dal* had been protecting the village *soyam* land for the past 20 years, paying a watchman with voluntary contributions. In Naurakh and Resal, civil land was being protected by individual families through private enclosures. Officially termed 'encroachment' on government lands, such informal systems are fairly widespread due, in part, to their low transaction costs (see Singh, 1997a, 1997b).

## Van panchayats

*Van panchayats* were created through notification of the Kumaon Panchayat Forest Rules in 1931 and have since undergone several changes. Notably, the rules were designed on the basis of existing traditional communal-property resource-management systems in the area and were *not* notified under the IFA, but under the Scheduled Districts Act of 1878, then applicable in Kumaon. This permitted development of special arrangements for the unique cultural and biophysical features of the mountainous region.[26] The original framework provided by the 1931 rules is summarized in Box 3.5.

---

**Box 3.5** *The Kumaon Panchayat Forest Rules, 1931*

Any two or more rights-holding residents of a village could apply to the deputy commissioner (DC) to demarcate a specific forest area within the village's 1823 boundary as a village forest provided that one third or more of the rights holders in that area did not object. After dealing with any claims or objections, the DC called a meeting of the residents and other rights holders for electing three to nine *panches* (members of the forest council) for managing the village forest. The *panches* selected a *sarpanch*[27] from among themselves.

 The elected representatives signed an agreement that the village forest land would not be sold or partitioned and that 'The produce of the *panchayat* forest shall be utilized by the *panchayat* to the best advantage of the village community and of the right holders'.[28] The *panchayat* had the status of a forest officer with the powers to fine or prosecute offenders and 'to sell forest produce,[29] including slates and stones, without detriment to the forest, and to issue permits and charge fees for grazing or cutting grass or collecting fuel'. Resin from Chir (pine) trees was the only product that could not be extracted or sold without the permission of the Forest Department, and resin income had to be shared with the department where extraction was done by the latter. *The van panchayat had full control over use of its income from the forest, and all dues payable to it were deemed as dues payable to the government and recoverable as arrears of land revenue.* The only role assigned to forest officers was to inspect and report on the *panchayat* forests or their records, if requested to do so by the deputy commissioner.

According to recent estimates, there were 6069 *van panchayats* managing 405,426 hectares of forests (13.63 per cent of total forest area) in Uttarakhand (FDUP, undated). Following revision of the Panchayat Forest Rules in 1976, these forests are now demarcated as village forests under section 28 of the IFA and are entered in the land records in the *panchayat's* name. *Thus, both the institution of the* van panchayat *and the village forests under its management are legally constituted. This is in contrast to the administrative orders governing the village institutions and forest lands brought under JFM in other states.*

Until 1976, *panchayati* forests could be carved out of both reserve and civil (protected) forests falling within the 1823 village boundaries. The last revenue settlement conducted during the early 1960s, however, redefined village boundaries, taking all Class I reserves out of them. These reserves were subsequently transferred back to the Forest Department, effectively blurring the distinction between Class I and Class II reserves. Revision of the *van panchayat* rules in 1976 made only the forests falling within the *new* revenue village boundaries eligible for conversion to *panchayati* forests. Consequently, the majority of subsequent *van panchayat* forests have been demarcated out of civil forest land under the revenue department's jurisdiction. The area under each *van panchayat* ranges from a fraction of a hectare to over 2000 hectares.

The development and growth of *van panchayats* can be divided into three major phases. The first phase lasted from 1931 until the early 1960s; the second lasted from the 1960s until the early 1990s; and the last one began from the mid-1990s, when there was a revival of state interest in them with the dawn of participatory forest management policies.

## *Van panchayat* functioning: Phase I (1931 to early 1960s)

The Kumaon Panchayat Forest Rules, 1931, were notified almost ten years after the recommendations of the Kumaon Grievances Committee to create *panchayat* forests. During this period, an open-access regime was created, and both the peasantry and the state engaged in uncontrolled extraction from Class I reserves. The Panchayat Forest Rules enabled the villagers to seek demarcation of specific forest areas for re-establishing communal management. However, these effectively replaced 'a system of customary rights, enforceable communally and also judicially, by the institution of the *van panchayat* which came into operation only *after* it was sanctioned individually for each village by the divisional commissioner who sat at Nainital' (Agarwal, 1996).

Only 428 *van panchayats* were formed by the time of independence and 1074 by 1960. Initially, two major factors inhibited villagers from applying for *van panchayats*. Firstly, the traumatic experience of forest reservation had made most villagers suspicious of the colonial administration. Secondly, only those villages with educated men could negotiate their way through the bureaucratic labyrinth when applying for a *van panchayat*.

By all accounts, however, these early *van panchayats* functioned well with limited, if any, Forest Department involvement in their affairs. They improved forest condition dramatically through effective exclusion of outsiders and

managed forests for livelihood and subsistence needs. In 1960, the Kumaun
Forests Fact-finding Committee found the condition of *panchayat* forests to be
'generally satisfactory' (GoUP 1960, p33). A study of 11 *van panchayats* in 5 of
the 8 hill districts during 1983 to 1984 by the state planning division found that
all of them had prevented illegal felling and damage due to fire; 10 had prevented undue damage to the trees; 9 had prevented encroachments; and 8 had
exploited forest produce scientifically. It also found that since the formation
of the *van panchayats*, forest wealth had increased by 40 to 50 per cent (GoUP
1984, p28). From a random sample survey of 21 *van panchayats* in Nainital,
Almora and Pithoragarh districts, Somanathan (1991) concluded that *van
panchayats* have, by and large, maintained oak forests very well, especially
in contrast to the dismal condition of the reserves near habitations, though
Chir forests seemed to have done as badly under *panchayat* control as in the
reserves.[30]

Many *van panchayats* with good income from resin and charcoal invested
in village development activities, such as building schools and hostels for
children (PSS, 1998). They received support from revenue department field
functionaries for resolving inter-village boundary disputes, obtaining maps
showing their forest boundaries and enforcing their authority against violators
of their rules. Although women remained absent from *panchayat* councils, there
was fairly democratic decision-making among the men. Our study of Dungri
Chopra Village (see Box 3.6) indicates the quality and resilience of early *van
panchayats* compared to today.

The other case study *van panchayats* also evolved diverse rules – adapted
in response to changing requirements – through collective decision-making.
*Van panchayats* enjoyed considerable autonomy in decision-making and
control over the forest. They balanced the maintenance of ecological services,
such as soil fertility and water source protection, with grazing, collecting and
other forest uses necessary to support local livelihoods. Our cases suggest that
high stakes in the forest and strong bonds of trust among villagers allowed
many of the *van panchayats* to remain successful for many years. The *van
panchayat* in Chora, for example, was formed in 1946. Until the introduction of
VFJM, the *van panchayat* functioned very well, despite lack of support from
revenue officials in enforcing its rules. It held regular meetings, working
closely with an active *mahila mangal dal* to protect the village forest and to
regulate collection of fuelwood, fodder and other forest products. Similarly,
the *van panchayat* in Gadyuda, formed in 1965, managed a pine forest for
several decades to generate income for village development activities. The *van
panchayat* in Makku, formed in 1958, managed 2200 hectares of forest, in which
85 villages continue to exercise legal rights. For the past ten years, a democratically elected *sarpanch* ensured regular general body meetings and participatory decision-making. He also encouraged several *mahila mangal dals* to
protect communal *gram panchayat* land closer to villages to better meet their
fuelwood and fodder needs, while also saving it from encroachment by the
elite. Both he and the women faced resistance from powerful vested interests

**Box 3.6** *The* van panchayat *of Dungri Chopra*

Dilip Singh of Dungri Chopra Village, who was instrumental in getting the *van panchayat* formed in 1939, said that village elders were firmly against the proposition. They feared that *van panchayats* were a ploy of the colonial government to snatch their village forest, leaving them with no area for grazing cattle or collecting grass. Memories of the *begar* (free-labour) system were still fresh and the village elders feared its return. Nevertheless, four to five younger men joined Dilip Singh to form a *van panchayat* for about 16 kilometres of forest and made Amar Singh, who was most resistant to the idea, the first *sarpanch*.

The condition of their forest had become deplorable after access to Class I reserves was opened to all *bona fide* residents of Kumaon. Only a few sal trees survived. The *van panchayats'* first major task was to reassert customary village authority over its forest. The divisional forest officer gave them a small amount of money to build a protection wall, which the *sarpanch* distributed equally among every man, woman and child who helped. Neighbouring villages, which had been using the forest during the years of open access, resisted its enclosure. The residents of Khobra Village even filed a case against Dungri Chopra. The case failed and the forest was totally closed for three years. Villagers planted *Banj* oak seedlings and pine seeds obtained for them by the divisional forest officer from Nainital. According to Dilip Singh, who remained the village *pradhan* for 30 years, things were very different in those days. There was unity in the village and disputes were resolved by the village *Nyaya* (justice) *Panchayat*. There were few government schemes and the villagers built the school and the *panchayat* building themselves.

Today, schemes worth *lakhs*[31] come to the villages and there is rampant corruption. No government official visits the village without negotiating a commission in advance. With most men having migrated away in search of jobs, male interest in managing the village forest has declined. A few years ago, the village women succeeded in getting an all-women *van panchayat* council elected. In 1999, the district rural development agency sanctioned 60,000 rupees for undertaking plantation in the village forest. When Dwarka Devi, the woman *sarpanch*, went to collect the first installment of 30,000 rupees from the *van panchayat* inspector, he made her sign a receipt for the full amount but gave her only 24,000 rupees. She went to Dilip Singh to seek advice on what to do. He told her that in future, whenever any such payment had to be collected, she should always take other women *panches* with her, and on returning to the village, she should place the entire amount in front of the general house to prevent anyone from suspecting her. The villagers would themselves help her work out how to deal with the situation. Dwarka Devi internalized this valuable lesson in transparent governance. This enabled her to maintain collective responsibility for managing the village forest and to evolve coping strategies for dealing with the increasingly unsavoury world outside of the village. The *panchayat* forest was one of the best in the district and the women met almost all their forest needs from it. They even permitted every household to harvest one timber tree each for their own needs a few years ago (Gairola, 1999b).

and husbands who had to do housework while women patrolled. Effective protection by the women led to dramatic regeneration of the women's forests.

Many villagers, however, could not negotiate the bureaucratic procedures for constituting a *van panchayat*. In other villages, people saw little advantage in constituting formal *van panchayats* as they continued to assert customary authority over their commons on the strength of the 1823 boundaries. Consequently, the total number of *van panchayats* remained low.

## The erosion of *van panchayat* authority: Phase II (1960s to early 1990s)

In 1956, the revenue department abolished the post of the divisional *van panchayat* officer, substantially slowing *panchayat*-related paper work and other support work (Singh, 1997c). In 1964, 3000 square kilometres of the Class I Reserve Forests were transferred back to the Forest Department from the Revenue Department.

The biggest blow for the *van panchayats* came with the revision of the 1931 rules in 1976. These drastically curtailed *panchayat* autonomy, authority and entitlements. The new rules restricted the area eligible for new *van panchayat* formation to sites falling within the new village boundaries, drawn under the last revenue settlement of the early 1960s, instead of within the 1823 boundaries. As the new village boundaries excluded Class I Reserve Forests, this amounted to a steep reduction in the forest area available for *van panchayat* control. *While the villagers continued to depend on these areas, they were no longer permitted to manage them* (Saxena, 1995b).

The revised rules also allocated 20 per cent of the *van panchayats'* forest income to the *zila parishad* (district-level government) for development works and 40 per cent to the Forest Department for re-investing in *panchayat* forests. The remaining 40 per cent share that was left for the *panchayats* could no longer be used without prior permission from the sub-divisional magistrate or deputy commissioner. This effectively deprived the *van panchayats* in remote villages from access to their own drastically reduced share of income, as the costs of repeated trips to distant offices outweighed the benefits. The revised rules made the Forest Department responsible for preparing working plans for all *panchayat* forests, thereby expanding its technical authority substantially.

There is no evidence that the Forest Department used its expanded authority to improve planning or to invest its share of income in *panchayat* forests. The revenue department similarly abused its expanded authority in many of our case sites. *Van panchayat* councils were suspended arbitrarily, with no fresh elections held for years at a time. Despite repeated requests, Bareth's *van panchayat* was not informed about the amount credited to its account for resin tapped three times from its forest since 1974. Requests for assistance in dealing with encroachments on *van panchayat* forests met with a stony silence, with some local revenue officials actually abetting the encroachers. In Gadholi, village youth were arrested for demanding the removal of encroachers colluding with the Forest Department, and the whole village protested in their support (SKS, 1999a). Even deputy commissioners responsible for preventing misuse of *panchayat* funds abused their authority, claiming the money represented achievement of district small savings targets.

The revised rules also allocated most decision-making and administrative responsibilities to the *sarpanch*, in contrast to the earlier focus on the *van panchayat* council. These responsibilities include calling and presiding over *panchayat* meetings; executing all works; maintaining accounts; supervising employees; maintaining all of the specified files; undertaking correspondence on behalf of the *panchayat*; and filing or defending court cases. For all of these responsibilities, individuals are entitled to spend the grand sum of 50 rupees, a sum not revised since 1976. This change increased administrative convenience but negatively influenced the *van panchayat*'s internal governance. Over the years, the administration has also issued a plethora of administrative orders regulating *panchayat* functioning. Each *sarpanch* is required to maintain as many as 17 files and stamps (Dubey et al, 2000), making the post an onerous responsibility, particularly as it carries no financial compensation.

In 1981, commercial felling in the hills was banned as a consequence of the *Chipko* Movement. Simultaneously, the Forest Development Corporation was given monopoly rights over salvage timber, even from *van panchayat* forests, which earlier was used by villagers for their own needs. The department stopped giving permits for bamboo and cane harvesting to artisanal producers during the mid-1980s. The Tree Preservation Order of 1976 deprived villagers of the right to cut trees even from their private lands without cumbersome Forest Department permissions. The latest threat has come in the form of the ever-expanding protected area network. Many *panchayats* fall within protected areas and villagers have lost all or most of their rights in both village and other surrounding forests.

These policies weakened *van panchayats'* abilities to manage forests for the benefit of their people. Boundaries marking panchayat authority were blurred, especially between *van panchayat* and civil/*soyam* lands. Confusion has been compounded by arbitrary allocation of the area of one village to another, and frequent redrawing of village and *gram panchayat* boundaries.[32] As boundaries define communal property rights, a very large number of *van panchayats* are embroiled in boundary disputes. Re-allocations have also created dramatic inequalities among villages in the kinds of forest resources that they can access. Progressive weakening of traditional inter-village dispute resolution mechanisms has trapped the *van panchayats* in prolonged and expensive litigation.

The concentration of powers and responsibilities in the *sarpanch* is also weakening collective decision-making by *van panchayats*, while reducing transparency and accountability. The bureaucracy has also attempted to cast small village institutions in its own mould by requiring each *van panchayat* to maintain as many as 17 different files/seals, bogging the *panchayats* down in administrative tasks. The FCA, the felling ban, monopoly control of the forest corporation on resin and even salvage timber, and controls over NTFP marketing have drastically reduced the livelihood and employment benefits from village forests, while *van panchayats* have been divested of direct control over the limited income that remains their entitlement.

In the words of N K Maithani, until recently, the able *sarpanch* of Makku Van Panchayat:

> *After the British took them over, the people never really regained control over their traditional forests. The FCA [Forest Conservation Act, 1980] and the felling ban in the early 1980s drastically reduced the viability of large village forests being maintained by* van panchayats.

Earlier, the *panchayat* raised funds for protection and management of its 2200-hectare forest by making charcoal with Forest Department permission. No such permission has been granted since 1980, so that it is a struggle to raise resources for managing the large village forest.

## New *van panchayats*: mid-1990s to present

*Van panchayats* survived these multiple obstacles and challenges to their authority in many villages, albeit in a weakened state. They have faced two additional threats recently, however, which may bring about their demise as relatively democratic self-governing institutions. Ironically, these threats are state initiatives presented as 'devolution' policies: the rapid formation of new *van panchayats* under the direction of the revenue department; and the introduction of village forest joint management (VFJM) by the Forest Department. Their practical effect has been to transfer authority still further to the state at the expense of communities.

The demarcation of remaining civil/*soyam* lands as village forests increased rapidly over the past few years. At the time of independence, there were only 61 *van panchayats* in Nainital District. By 1995, another 138 had been constituted. In December 1999, Nainital technically became the first district in the country in which all of its 495 villages with at least some civil forest land had a legally demarcated village forest (Dubey et al, 2000). It is too early to tell what the effects of state-driven *van panchayat* formation might be on local livelihoods, social relations and forest management practices. The process of their formation, however, gives reason for concern.

## *Organizational inclusiveness*

Demarcation of village forests by the revenue department has been a supply-driven, rather than a demand-driven, process. Instead of the villagers collectively deciding to apply for a village forest, the administration decided that they should have one, irrespective of ongoing boundary disputes, existing community management arrangements, land shortages, or encroachments or land uses developed over years of neglect. In Resal Village, residents were kept in the dark about the content of papers that they were asked to sign – papers that authorized conversion of their civil land into a village forest. When villagers found out, they were furious, and demanded that the deputy commissioner return their 'application' as they have an established system of civil land management through household enclosures. They were particularly suspicious of state intentions because the forest surrounding their village has been declared a wildlife sanctuary. The Wild Life wing of the Forest Department also attempted to make them form an eco-development committee and develop alternatives to their forest-based livelihoods. Villagers rejected this outright as they are agitating for the restoration of their forest rights.

## Inter-village relations

Some multi-village *van panchayats* have been reorganized in ways that exacerbate inter-village conflicts. In Anarpa, for example, villagers developed an effective multi-village governance system that was both democratic and equitable. Without any say in the matter, they were told that their multi-village *van panchayat* would be divided, destroying existing mechanisms for sharing resources and leading to major disruption and anger (see Box 3.7).

---

**Box 3.7** *Anarpa* van panchayat: *impact of top-down reorganization*

Anarpa had an old *van panchayat* that was formed during 1945 to 1947 and that covered four villages. The villagers depended on the forest for fodder, grass, *kafal*, *amla*, branches for supporting vegetable plants and animal bedding. Earlier, the *van panchayat* also earned income from resin and timber. Forest dependence had increased recently as requirements of fodder and branches for vegetable cultivation could no longer be met from cultivated areas. The *van panchayat* forest was adequate for meeting all four villages' needs; the villagers did not need to go to the reserve forest.

The *van panchayat* had fairly equitable and enforceable rules. These included differential fees for firewood and fodder for *panchayat* residents and those of three neighbouring villages, and grass collection permits for only one person per household. Everyone went to put out a fire when necessary. The majority of the households are scheduled castes, including the *sarpanch*, for the ten years prior to reorganization. The council had two women who voted but who otherwise did not participate in *panchayat* meetings.

Suddenly, in 1999, revenue department staff divided the *van panchayat* into four separate *panchayats*: one for each village or hamlet. As neither the forest area nor the species composition could be evenly distributed among the four villages, some are now left with only small patches of Chir pine, while others have all the fodder-bearing areas. This has created a potential for boundary disputes where none existed before. The villagers resented this new arrangement. The women suffer worst as they now have to contend with the watchers of four *panchayats* while collecting forest products (SKS, 1999b).

While new *van panchayats* exclude the most forest-dependent villagers from decision-making and threaten inter-village relations, they also fail to address any of the major problems that plague existing *van panchayats*: the lack of effective and easily accessible dispute resolution mechanisms; inter-village inequity in the availability of forest areas; and erosion of *panchayat* authority. The rapid formation of *van panchayats*, rather than expanding space for local forest management, seemed to be reducing it still further.

## Village Forest Joint Management (VFJM)

VFJM, implemented by the Forest Department, was even more problematic for villagers. UP Forest Department's innovation with respect to JFM was to bring *van panchayat* and civil/*soyam* lands within the Forest Department's purview through the Village Forests Joint Management Rules, 1997 (see Box 3.8). Whereas in most other states JFM enables villagers to participate in managing forest lands under the Forest Department's jurisdiction, in Uttarakhand, the VFJM rules enable the department to become the dominant partner in the management of *village* forests. Much of the land being brought under VFJM falls under the jurisdiction of *van panchayats* or the revenue department/*gram panchayats*, and *not* the Forest Department.[32]

The VFJM rules also provide for the formation of village forest committees under the UP Panchayati Raj Act in villages where there were no *van panchayats*. This was an effort to link VFJM with the *gram panchayat*, the lowest institution of local self-government. These forest committees were expected to be representative of key local interests, with one seat each designated for women, scheduled castes/tribes, backward castes, and for persons with a particular interest in forests. The *pradhan* of the *gram panchayat* was to be the president of the forest committee and the forest guard its member secretary. While the objective was desirable, the order was again a top-down prescription. It did not strengthen participatory governance by the *gram panchayats* and forest committees or enable forest committees to benefit from their link with local government institutions. Rather than devolving greater authority to self-governing village institutions, VFJM extended the forest bureaucracy's jurisdiction and technical supervision even to civil/*soyam* and village forest lands.

Notification of the VFJM rules, together with other orders for 'participatory' forestry, was a condition of the World Bank US$65 million loan for the Uttar Pradesh Forestry Project over the period of 1998 to 2002. JFM was to receive priority under the project, accounting for about 30 per cent of the total budget. The World Bank's appraisal document did not provide any analysis supporting the introduction of VFJM, instead of strengthening existing *van panchayats* and informal community institutions. Nor did it specify any process ensuring stakeholder participation in framing the 'participatory' orders. The World Bank project simply imported the standard JFM model from other states, with all of its shortcomings.

---

**Box 3.8** *Devolution or centralization? The language of VFJM rules, 1997*[34]

Rule 3(1) commences with '*Subject to the supervision, direction, control and concurrence of the Divisional Forest Officer,*[35] a village forest shall be managed jointly by the 'Village Forest Committee' and such officers of the Forest Department as are nominated in this behalf by the Divisional Forest Officer, on the terms and conditions specified in Form 1 [containing the agreement to be signed by the village forest committee (VFC)/ *van panchayat* (VP) with the governor of Uttar Pradesh (UP)]'.

Rule 3(2) states: 'If a forest *panchayat*, by resolution, decides that *panchayati* forest under its management be managed in accordance with these rules, *the provisions of the Uttar Pradesh Panchayati Forest Rules, 1976, shall cease to apply in respect thereof.*' This provision generated tremendous confusion among Forest Department staff and villagers, as nothing was said about what was to be done with existing accounts, funds and activities of the *van panchayat*.

The Uttar Pradesh Village Forests Joint Management Rules, 1997 (VFJMR, 1997) use the generic term 'village committee' for both village forest committees and the *van panchayats*. During fieldwork, it was found that *van panchayats* participating in VFJM were increasingly being referred to as village forest committees, confusing their very identity.[36]

The functions and duties of the village forest committees specified in the rules include those specified in the 1976 Van Panchayat Rules, such as protecting trees and preventing encroachment, in addition to '*abiding by the approved micro-plan and carrying out the directions of the Divisional Forest Officer*'.

Clause 5 of the agreement provides 'that if the beneficiary fails to carry out any of the directions issued by the Forest Officer for forest management or any of its binding obligations, the *Forest Officer shall carry out at his discretion any or all of the works regarding forest management departmentally*'. The Forest Department itself did not fulfill any of its responsibilities towards *van panchayats* under the 1976 rules. Nevertheless, it was again assumed that only villagers would fail and no mechanism was designed to increase the department's accountability for honouring *its* responsibilities. Furthermore, the conservator of forests has been empowered to be the final and binding arbitrator in disputes between the two parties.

---

## The impact of VFJM on communities in Uttarakhand

It is too early to perceive the long-term impacts of VFJM on forest-based livelihoods and forest quality. The content of the VFJM rules, however, suggests a loss of decision-making space for local villagers. In the following sub-sections, we highlight a few early effects before discussing how forest users were reacting and why.

### *Organizational inclusiveness*

The Forest Department prioritizes and selects villages for VFJM in accordance with several criteria. 'Spearhead teams' communicate with, and develop micro-plans for, selected villages. These teams consist of: one assistant conserv-ator of forests; one ranger or deputy ranger; one forester or forest guard; and two NGO 'social motivators', at least one of whom should be a woman (see Box 3.9 for a discussion of NGO roles in VFJM). The agreement to be signed by the participating villages refers to them as *beneficiaries* rather than as right-holding partners. The ex-*sarpanch* of Makku Van Panchayat found this term highly offensive, saying that 'the villagers are legal forest right-holders and not beneficiaries receiving doles from the Forest Department'.

As for the new *van panchayats*, new village forest committee formation was a supply-driven rather than a demand-driven process. Even where *van panchayats* had been in existence, small groups of elite men, with the least dependence on the forest, often made alliances with Forest Department field staff to subvert the requirement of obtaining *informed* general body resolutions that accept VFJM. The villagers' role was reduced to providing information to spearhead teams during rapid appraisal exercises. The plans were actually written by Forest Department staff, in the mould of the department's annual plantation projects, and subsequently implemented with the involvement of only the *sarpanch* and/or some of the elected *panchayat* members, or by the *pradhan* of the *gram panchayat*. Villagers provided only wage labour.

Since *gram panchayats* covered more than one village, the village forest committee president was not necessarily even a resident of the village that managed its forest.[37] Existing community management systems, including the relationships established by them with their *gram panchayats*, were over-ridden with potentially damaging long-term impacts. Women's groups, in particular, had been able to negotiate their authority to manage civil/*soyam* lands with their *gram panchayats*. In at least three out of ten case studies (Pakhi, Arakot and Chora) where VFJM was introduced, village women were actively pro-tecting the *van panchayat*/*soyam* forests. In all three cases, no effort had been made to build upon and strengthen the women's efforts.[38]

## Democratic leadership

In our case studies, after forest committees were constituted, there was no discussion of their affairs in *gram panchayat* meetings, making the link between the two notional rather than real. Once information for micro-planning was collected, the Forest Department staff interacted primarily with the *sarpanch* or *pradhan* in order to get him or her to execute the planned activities. Even where general body or *van panchayat* council meetings were regularly held, as in our case study in Chora, the meetings declined in frequency or stopped following the introduction of VFJM. In the majority of cases, ordinary villagers and even *van panchayat* council members had little information about what was happening with regard to the management of their forests, and most perceived a shift in control over their *van panchayat* or civil/*soyam* forests to the Forest Department.

---

**Box 3.9** *NGOs as service providers in VFJM*

Although spearhead teams are presented as a unique and successful innovation of the Uttar Pradesh Forestry Project (FDUP, undated), the arrangement placed non-governmental organization (NGO) staff in a highly compromised position. The assistant conservator of forests evaluates NGO team members and can withhold their payments on the grounds of unsatisfactory work. A woman staff member of a leading NGO said that although the initial three-week training had generated enthusiasm and high expectations, the actual functioning of spearhead teams was *ad hoc*. They spent one or two days preparing each micro-plan instead of the required three weeks or more. She often received calls at short notice to reach a particular village where she had not worked before for a VFJM activity. She felt that the arrangement was a ploy to destroy the credibility of NGOs. The Forest Department staff, on the other hand, complained of multiple duties and inadequate time for genuinely participatory micro-planning. Senior officials often over-ruled field staff decisions, making it difficult to build credibility with villagers.

In all but one of our VFJM case study villages, there were serious problems with NGO work. In Naurakh, the NGO representative accompanied Forest Department staff late one evening and asked only a handful of village men to hurriedly pass a back-dated resolution accepting an already prepared micro-plan. In Pakhi, where village women active in the *Chipko* Movement were managing the *van panchayat's* forest, the spearhead team reached an agreement with the men without even informing the women. In Arakot Village, although a *mahila mangal dal* was protecting the forest for 20 years, a male majority forest committee was formed.

In Uttarakhand, NGOs and civil society groups have historically played a strong advocacy role. *Chipko,* for example, was triggered by protests led by *Dasholi Gram Swaraj Mandal.* Today, the NGO movement is split into different camps and factions. The vast majority have been co-opted to work as 'private service providers' for several donor-funded projects, including the forestry project. Once they accept working on project terms, they effectively lose their critical and questioning voice. Consequently, no unified voice was raised against the potential damage to the region's unique *van panchayats* from VFJM. A large number of concerned individuals and advocacy groups, however, articulated such concerns at different forums (SKS, 1999c; SPWD, 2000).

## Control over funds

While the World Bank loan required revision of the 1976 Van Panchayat Rules to provide them with 'greater authority and control of funds', ironically, the 1997 VFJM rules and the revised Van Panchayat Rules of 2001 did exactly the opposite. The latter provide for the *van panchayat's* accumulated income, lying locked up with the district commissioners, to be transferred to *panchayat* common funds, in order to improve accessibility. The future share of forest income is also to be increased from 40 per cent (under the 1976 rules) to 80 per cent. However, villagers' access to their income may not improve substantially, as a Forest Department functionary will now be placed *inside* each *panchayat*, controlling its day-to-day activities as joint account holder and member secretary, in contrast to earlier control over only *monetary* income by a distant divisional commissioner. The *panchayat's* proposals for using their income will require prior approval by the divisional forest officer (instead of the district commissioner) who may modify them if she or he deems fit (GoU, 2001). Foresters thus enjoy final say over how the village institutions use their own income. A question repeatedly asked by local activists was why a loan from the World Bank was required for undertaking plantations on *van panchayat* forests when such large sums of their existing income (lying with deputy commissioners) could be released for the purpose (SPWD, 2000)? The villagers' unhappy experience with such an arrangement under JFM in Harda Forest Division in MP has been mentioned in the earlier section 'Madhya Pradesh'.

## Livelihoods and equity

A key assumption of VFJM is that the major problem plaguing *van panchayats* is lack of funds, and the best incentive for increasing villagers' stake in forest protection is to offer them attractive shares of income from sale of forest products. Yet, a survey of 644 *van panchayats* in Ranikhet Sub-division in Almora District found that as many as 433 did not have any income, and only 45 could boast of a balance of at least 25,000 rupees in their passbooks (Singh and Ballabh, 1991, cited in Singh, 1997c). A large number of forests were managed well by villagers without any source of income. It has also been observed that the income of a *van panchayat* has no bearing on what the villagers consider to be a 'good' *van panchayat*. A *panchayat* was often regarded as good if it met fuel and fodder needs and helped to recharge water sources. Oak forests were generally preferred to pine forests, even though they provided less revenue and employment (through resin-tapping). In contrast, the revenue department considered *van panchayats* to be performing better if they had bigger bank balances (Singh, 1997a). The World Bank forestry project subscribed to the same assumption. Discussions on the merits and demerits of VFJM rules often centre on the percentage of the share of *income* that the villagers would get from their forests. Women forest users, however, have attacked one another while competing for increasingly scarce fuel and fodder resources, some even resorting to suicides to end their daily drudgery (Nanda, 1999). Their priorities are to increase the direct *use* values of their forests, although additional income is never unwelcome. The project document claims to target women and the poor, but provides no analysis of how a shift in management priorities to increase income would impact their access to daily subsistence or their work burdens. VFJM micro-plans in the case studies had few provisions for supporting livestock, a critical local livelihood requirement; instead, they emphasized closure of forests to grazing.

## Employment and other benefits

The World Bank-funded forestry project provided an average of 1.5–2 million rupees for implementing a micro-plan in each village under VFJM. This promoted inequity among villages. The sudden offer of large sums of money to selected villages with high unemployment and limited opportunities for cash incomes has also led to the eruption of major conflicts within villages to gain control over the funds. Sustainable voluntary protection, often by women's groups, had been replaced by patrols of externally funded watchmen. In our case study villages, male elite selected paid watchers, providing a new avenue for patronage. In Pakhi Village, a poor watchwoman, who was paid by women's voluntary contributions, was replaced by four watchmen, who were paid much higher salaries with project funds.

While overlooking existing systems of voluntary contributions, the project demanded that villagers contribute 20 per cent of micro-plan costs. Contributions were collected through compulsory deductions from wages, transferring their costs to those performing wage work – primarily women or poorer villagers. In none of the case study villages had any open discussion been held on how the mandatory contribution could be shared more equitably by all of those theoretically gaining entitlements to the specified benefits. In Kharag Karki Village, women thought they could at least take the firewood from cleaning operations as compensation for accepting lower wages; but they were not permitted to do so. This left them alienated and angry. Acutely forest-dependent women bore disproportionate costs of (in)voluntary contributions and unpaid protection duties in order to build up *panchayat* and forest committee funds controlled by the male elite.

## Seizure of the best village forests through VFJM

Despite villager demands to extend JFM to well-stocked state-owned forests, the forest departments of most states have restricted it to only degraded forest lands (the section on 'Madhya Pradesh' provides an example of an exception). In Uttarakhand, however, better-quality *van panchayat* forests were being selected for VFJM. This actually enabled the Forest Department to claim credit for years of management effort by the villagers. Several NGOs have questioned the justification for bringing well-protected *van panchayat* forests under VFJM when large areas of reserve forests under the department's exclusive jurisdiction are acutely degraded and could be improved through JFM (SPWD, 2000).

## Resistance to centralizing devolution

Efforts had been made to introduce VFJM in ten of our case study sites. Seven of these had existing *van panchayats* and two others had informal community management systems in place. VFJM was rejected outright in two sites, one with an established *van panchayat* (Makku) and another (Naurakh) where villagers had been protecting their civil forest land for four to five decades through private enclosures. In two other sites, there was no *van panchayat* and, consequently, village forest committees were formed under the UP Panchayati Raj Act. Thus, VFJM had been started with existing and functioning *van panchayats* in six sites.

## Why did *van panchayats* accept VFJM?

Given the assertion of Forest Department authority on *panchayat* forests implied by the VFJMR, 1997, and the kind of impacts discussed above, we tried to understand why villagers agreed to VFJM. One of the most important reasons was a lack of information. In none of the six villages with existing *van panchayats* were the majority of villagers aware of the provisions of the VFJM rules. In Pakhi, where the women's association had effectively been managing the *van panchayat* forest, the women did not find out that some men had accepted VFJM until after the agreement had been signed. At best, the *sarpanch* and, sometimes, the other *panches* had received copies of the rules. Such leaders, however, had either not studied them carefully or made no effort to obtain informed consent for VFJM from the women and men of the community. The majority of villagers perceived VFJM to be another of the many government plantation schemes that they have seen over the last two to three decades, offering short-term wage employment. Even the elected *panches*, other than the *sarpanch*, knew little about the content of the agreement that had been signed with the Forest Department suspending the *van panchayat's* autonomy. The *sarpanch* of Cheerapani wanted to cancel the agreement after his relations with department staff became strained, but was told that the rules had no such provision. In Arakot, there was almost panic among the villagers, particularly the women, on learning of the VFJM agreement. They said that they would ask the divisional forest officer to sign another agreement, assuring them of continued control over their *soyam* land. An ex-*sarpanch* of Gadyuda was concerned about how the village would repay the loan with which the plantation work was being undertaken. When he asked the divisional forest officer, he received the retort: 'Would you like us to stop this scheme in your village?' A woman *panch* of Chora complained during a VFJM workshop that the villagers had no information about the expenditure on their micro-plan. She was told that 'If you don't know anything, then why have you come?'

Only information about the micro-plan budgets had been selectively publicized. This had kept ordinary village women and men off guard about the transfer of control over their village forests to the Forest Department. The manner in which the resolution for VFJM was passed by Chora Van Panchayat, with a record of regular meetings, diverse rules and an active *mahila mangal dal* protecting the forest, is illustrative (see Box 3.10).

---

**Box 3.10** *Introduction and impact of VFJM on Chora Van Panchayat*

The resolution for VFJM in Chora was passed on 27 July 1998 and has only four signatures on it. The second page has the signatures of 19 women. According to *van panchayat* council members, the forest guard came one evening and said: 'Hurry up, the money is about to come.' The meeting was held the same evening with the available *panches*. The guard told them that they could get signatures of more people later. The *panches* got the women's signatures the next day.

Custody of all of the *van panchayat* records, including the micro-plan, had now been grabbed by the deputy head of the *gram panchayat*. He did not permit the researchers to organize a village meeting to discuss issues related to VFJM raised by members of the women's association. He screamed at them, saying that 'Even if the DFO [district forest officer] comes, the records will stay with me. I am the one who brought this scheme to the village. I even went to Nainital for the purpose.'

The micro-plan had been prepared with considerable effort by the Forest Department staff with the villagers' role being confined to providing information. None of the ordinary villagers had seen it and knew little about its contents. A few villagers were aware of the total budget. The *mahila mangal dal*, which had been protecting the village forest voluntarily, was uninformed about the investments under VFJM. One of the women said: 'Earlier I used to attend the *panchayat* meetings. But now no general body meetings are held' (SKS, 1999d).

---

## Why was VFJM rejected?

In contrast to the cases above, in two instances villagers collectively rejected VFJM. Access to information about the content of the VFJM rules, and an opportunity to discuss their pros and cons among the village men (in the case of Makku) and among the village women (in the case of Naurakh; see Box 3.11) affected their decision to reject VFJM. In both cases, villagers' relations with the Forest Department were strained and the field staff were unable to provide satisfactory answers to their questions.

In the case of Makku, the large size of the multi-village forest vests the *van panchayat* with a status higher than even the *gram panchayat*. Elections for the *van panchayat* council are hotly contested and there is active participation by village men in frequent general body meetings. Progressive erosion of the *van panchayat's* entitlements to forest produce has reduced its substantial income of earlier days, making it suspicious of both revenue and Forest Department intentions. There was heated discussion over the pros and cons of VFJM in well-attended village meetings over a period of six months before it was finally rejected by the villagers. The range forest officer admitted that the Forest Department staff had been highly insensitive while approaching the villagers and unable to provide satisfactory answers to the villagers' penetrating questions about the VFJM rules. Rumours of some *van panchayat* land being included in a wildlife sanctuary doubled the villagers' suspicions about the department's intentions.

**Box 3.11** *Why the women of Naurakh rejected VFJM*

Naurakh is a village in Chamoli District that does not have a *van panchayat*. The majority of village households have long-standing private enclosures on village civil land. The guard and non-governmental organization (NGO) member of the spearhead team lured a handful of powerful village men into signing a back-dated resolution in favour of village forest joint management (VFJM), saying that a scheme with a budget of 2 million rupees had been approved for the village and that the resolution was urgently needed to prevent the village from being by-passed. The villagers, however, were explained the provisions of VFJM rules by a local youth activist with the help of a 'social army' of village boys and girls whom he had mobilized. This access to detailed information about VFJM terms, especially the potential loss of control over civil forests to the forest department, made the women vehemently reject it. They argued that, for them, retaining control over their civil land to meet fuelwood and fodder requirements was worth many times more than the 2 million rupees scheme. They already faced daily harassment by the forest department while collecting fuel and fodder from their civil land. They felt that they would not have any say in the selection of forest committee members, who would eat up the money. No amount of pleading by the village men in favour of wage employment could make the women budge (Gairola, 1999c).

## Summary and conclusions

Despite the imposition of crippling bureaucratic controls, a large number of Uttarakhand's *van panchayats* have survived as vibrant self-governing community forestry institutions. Where livelihood and ecological dependence on forests remains high, the *van panchayats* have successfully retained reasonable control over decision-making and the satisfaction of subsistence needs. The quality of forests under their management is often as good, if not better, than the reserve forests under the control of the substantially better resourced Forest Department.

Diverse, informal institutional arrangements for community management exist on all legal categories of forest lands, many led by acutely forest-dependent women. These co-exist with, and even within, the formal *van panchayats*. Such informal arrangements, often with negotiated support of elected *gram panchayats*, provide more accessible space for CFM to poor women and marginalized groups because they exist outside of the framework of bureaucratic procedures and controls.

Government actions substantially weakened these existing management systems over the course of the last century. Progressive restrictions on local livelihood uses of forest resources through the FCA, the felling ban and recent Supreme Court judgements, combined with large-scale livelihood and resource displacement caused by expansion of the protected area network, are changing people's attitudes towards forests and undermining the primary incentives for

CFM. *Ad hoc* changes in village boundaries and poor boundary demarcation of village forests, inequitable distribution of forests among villages, and inattention to conflict management among villages has soured inter-village relations. Classification of natural grasslands as forests under the management of a tree-focused Forest Department, together with extensive replacement of mixed forests by commercial pine plantations, has significantly changed land use and the nature of the forest itself over the years. The resource base available for sustaining Uttarakhand's agro-pastoral rural livelihood systems has diminished greatly.

State-driven devolution policies, in the form of new *van panchayats* and VFJM, fail to address these problems. Worse, they often undermine the remaining strengths of local forest management systems.

In the name of devolution, VFJM is empowering the Forest Department to reassert control over both *van panchayat* forests and civil/*soyam* lands, the only surviving village commons. Instead of revitalizing the rich and diverse base of indigenous knowledge of local women and men, and the management systems that they have developed for supporting livelihoods and maintaining ecological services, VFJM reinforces the Forest Department's claim to be the monopoly holder of technical forestry knowledge. *Van panchayats* and community management systems have survived *despite* such government interference and controls during the past. Users continue to resist intrusions by the revenue and forest departments through non-cooperation and withdrawal. There are, however, declining incentives for community management in a changing economic and social context and policy environment. Devolution policies such as VFJM may well lead to their ultimate demise.

## CONCLUSION

### Discouraging trends

State-directed devolution policies may appear to be progressive in geographic areas where state control over forests has been firm during recent decades, such as in the higher-quality reserve forests. There are few areas of this nature, however, among the villages now targeted for JFM. Our findings suggest, to the contrary, that for the three study areas, these policies have often reduced existing both *de jure* and *de facto* local space for forest management. The striking commonality in the JFM (VFJM, in Uttarakhand) frameworks of the three states is their a-historical nature. They overlook the diversity of *existing legal rights and land use* and location-specific, community systems for forest management often already in place.

Government agencies have not engaged villagers already involved with CFM in policy formulation, nor have they aimed to strengthen existing initiatives and institutions through jointly analysing local problems, perspectives and priorities. Instead, in all three states, JFM assumed that villagers need to be 'motivated' in order to protect forests in accordance with the Forest

Department's vision, even when this implies the destruction of the existing livelihoods of the poorest. The programmes have assumed that local people require monetary (in two states, with the help of World Bank loans) or timber incentives to 'participate' in state-defined management priorities. In the process, institutional arrangements producing relatively more accountable community leadership were replaced by male village elites, who were interested in increasing their personal power by allying with department staff. *Instead of active subjects negotiating devolution policy frameworks and their adaptation, villagers are treated as objects, to be reshaped by top-down policies into instruments to achieve externally defined resource management objectives and priorities.* As demonstrated by our cases, existing institutional arrangements that are more locally legitimate, more democratically accountable, and more sensitive to local livelihood needs, although not always equitable and gender sensitive, are being replaced by institutions that are amenable to extending state control.

Thus, state presence *inside* formerly autonomous community institutions is being engineered in all three states by imposing the forest guard or the forester as the member secretary of community organizations. In Uttarakhand and MP, this has allowed the state to control even community funds, including those raised by villagers through voluntary contributions, and to put them to uses favoured by foresters, often without any transparency or accountability. Existing social capital is either destabilized or destroyed as bureaucratic and non-democratic mechanisms for monitoring and controlling community behaviour are put in place as part of JFM agreements. These are the same non-transparent and non-accountable bureaucratic mechanisms whose failure, in the past, led to pressures for greater devolution and decentralization. *Rather than enhancing space for local forest management, such controls undermine faith in local communities as the custodians of local resources. While bureaucratic structures and controls are being dismantled for multinational capital in a globalizing market-driven economy, they are actually being expanded for the poorest forest-dependent communities in the guise of devolution.*

All of this takes place in the context of rapidly changing and contradictory legislation and public policy that legitimates central control and that state agencies can manipulate to suit their convenience. Foresters can pull out dusty maps with no relevance to existing land-use practices and claim authority to resources long used and managed by local people. Grazing areas used for many generations by villagers can become the targets of reforestation or closure to grazing on the strength of sweeping notifications issued during the past, declaring lands under a wide diversity of uses and ecosystems as state-owned 'forests'. Many of these lands have still not been surveyed or their long-standing uses or users properly recorded. Existing laws providing for devolution of their management to local government institutions, such as the Orissa Grama Panchayat Act of 1965 – vesting management of village common lands as well as protected forests in *gram panchayats* and *gram sabhas* – are ignored. Instead, the forest departments harp on the necessity of achieving a desirable percentage of 'forest cover', defined *ad hoc*, through ostensible devolution of forest management.

In all three of our study states, the highest 'space' for local forest management was available on lands where state presence was the weakest, which enabled CFM initiatives to flourish. Without acknowledging contestation of jurisdiction over common lands by revenue departments and local communities, state forest departments are consolidating their control in the name of devolution with both government and donor support.

This is aided by the growing and relatively uncontested environmental ideology of the urban middle class (with no direct dependence on forests), reflected in a sweeping Supreme Court judgement of December 1996 on a public interest litigation against illegal forest fellings. The court judgement brought *all* of the country's 'forests' (according to dictionary definition) on all legal categories of land (including private and community owned) under its ambit and banned all tree felling unless it was in accordance with technical working plans prepared by forest departments and approved by the central MOEF. In February 2000, the Supreme Court decided on another public-interest litigation case, banning all collection of NTFPs from legally notified protected areas, including the removal of dead and fallen trees within them. The impact of both of these, and other similar judicial pronouncements, on the livelihoods and land and resource rights of millions of already marginalized forest-dependent women and men has been drastic. The fact that these judgements overwrite other existing constitutional provisions and laws has gone relatively unchallenged due to the fact that the individuals who are most negatively affected by them are among the most disempowered and resource poor. There could hardly be space for further centralization of control over forest management in the forest bureaucracy. Discussion of the 'nitty grittys' of so-called devolution policies carry little meaning without clarification of what constitutes a 'forest' and the process by which that is defined.

The latest threat is looming in the form of rapid removal of all legal hurdles for attracting national and multinational capital in a fiercely competitive global market. The central government has already drafted a proposed amendment to Schedule V of the constitution of India, under which PESA was enacted, to legalize the transfer of tribal lands to non-tribals (including private industry) banned under the schedule. The central Ministry of Mines has similarly framed a policy for opening up forest and predominantly tribal areas for the mining industry. The empowering provisions of PESA, largely left unimplemented until the time of our research, will be left with even less meaning if these changes come into force. Rather than empowering forest-dependent communities in Schedule V (tribal majority) areas to bring their forests under community management, the proposed legislative changes will increase their already high vulnerability to displacement through increased acquisition of even their private cultivable lands for commercial exploitation. The nature of state response to organized tribal resistance against further displacement for 'development' became evident in December 2000, when the police shot three tribals dead in Maikanch Village in Rayagada District, Orissa. The objective was to break their determined refusal to permit surveys of their land for acquisition and leasing to a multinational bauxite mining company. Many of the villages to be displaced had held *gram sabha* meetings under PESA and passed resolutions against the proposed acquisition of their lands. In case of displacement, many of the villagers would not legally be entitled to any compensation as their lands were declared state-owned revenue wastelands or forests without any surveys recording their existence. The state response was to send in armed police to terrorize, frame false charges and kill (Das, 2002).

Finally, despite the increased rhetoric of participation, policy formulation itself remains non-participatory, generally deaf to the demands articulated by forest-dependent communities, CFM groups and their federations, and peoples' and civil society movements. Instead, senior government or Forest Department officials, with little understanding of forest-based livelihoods or direct experience of grassroots realities, enjoy the monopoly of formulating 'participatory' or 'devolution' policies. The intended 'beneficiaries' of supply-driven devolution policies remain deprived of the opportunity to shape the parameters of their own participation in forest management.

## Local appropriation of space for forest management: different motivations, greater impact

State forest departments, all the same, have been neither omnipresent nor omnipotent in nationalized forest management. Despite efforts to bring more and more land under Forest Department control, local users have been able to maintain or establish their own organizations for managing local forests due to the continuing importance of forests for livelihoods and the hardships caused by resource scarcities. Some of these organizations are rooted in cultural traditions that have survived two centuries of state interventions, while others date back to the early 20th century, including the legally constituted *van panchayats* in Uttarakhand. Many others are of more recent origin. Thousands of such self-initiated forest protection groups are protecting several hundred thousand hectares of state-owned forests in Orissa, Jharkhand, MP, and Uttarakhand, and, on a smaller scale, in several other states. Forests in these areas have been maintained in good condition over decades or have regenerated substantially over the past 10 to 20 years of recent local protection, and can now better serve a broad range of local livelihood needs. The forest protection committees of West Bengal (a large number of which represent formalization of self-initiated management efforts through JFM) received the Paul Getty Award during the mid-1990s, based on improvements in forest condition evident through satellite imagery. In Orissa, also, the Forest Survey of India recorded a 100 square kilometre increase in dense forest cover in just two districts between 1997 and 1999 as a result of CFM, some of which may have been formalized into JFM (FSI, 2000, pp82–83). Many village forests under *van panchayat* management in Uttarakhand are in as good, if not better, condition than reserve forests. The scale of impact of unofficial CFM in other states is less clear due to the absence of relevant data.

The more organized and better informed forest-dependent communities have also tried to take advantage of the provision for democratic decentralization under PESA in order to establish their authority over natural resource management in a holistic way. In pockets of MP, where tribal communities were mobilized and made aware of their rights under PESA by *Bharat Jan Andolan*, some communities have asserted control over all of their local natural resources, challenging the authority of the forest and other departments. In some of these villages, they have rejected the timber-sharing formula of JFM, preferring to manage their forests entirely for meeting local requirements (Behar and Bhogal, 2000; Sundar, 2000b).

In Orissa, thousands of self-initiated CFM groups have been building alliances by developing their own federations to articulate collective demands and to pressure the state government to formulate an alternative, enabling community forest management policy. They have challenged the state's JFM framework, pointing out the imbalance in power, accountability and entitlements between villagers and the Forest Department that it perpetuates. Civil society advocacy groups and NGOs that support these formations are facilitating the development of such an alternative policy framework, including a concept of community trusteeship rights (Singh, 2001) and evolving strategies for influencing the state government to replace JFM by a CFM policy.

Local organizations have also fought against donor initiatives that threaten local livelihoods, land rights and management authority. In MP, strong opposition to the Madhya Pradesh Forestry Project by an alliance of mass tribal organizations (MTOs) and people's social movements has made the World Bank wary of funding a second phase of the project, evaluated to be highly successful by the Forest Department and World Bank consultants. The grassroots organizations have challenged the forestry project's 'unsubstantiated assumptions about indigenous resource use and its supposed negative impact on forests and wildlife' (MTOs, 1999), and protested against not being involved during the phase of project formulation itself. Shaken by these protests, and under pressure from the World Bank, the state government and the Forest Department invited public inputs for developing a new state forest policy during 2000.

Lingering doubts about the MP government and the Forest Department's sincerity in initiating an open policy dialogue, however, were confirmed in April, 2001. Premediated state repression was unleashed on *Adivasi Morcha Sangathan,* a mass tribal organization in Dewas District of MP fighting for the local tribals' democratic rights. Members of state-promoted JFM committees helped to crush the tribals' democratic struggle for their customary forest and development rights in this first incident of its kind. On 2 April 2001, four tribals were shot dead by the police in Mehndikhera Village of Baghli tehsil during a sustained attack on several villages spread over many days, led by the district collector, the superintendent of police and the divisional forest officer. Members of 'official' JFM committees, many of them residents of the villages under attack, participated as paid labourers. Large numbers of houses, grain storage bins and agricultural implements were destroyed, sympathizers of *Adivasi*

*Morcha Sangathan* were physically assaulted and their women molested, and food grain stocks and drinking water sources were poisoned. And this was done in the context of a third successive year of acute drought in the area. The official version justified this action by stating that 'valuable' timber illegally felled by the tribals had to be recovered and that the tribals had to be prevented from 'encroaching' on state forests. Independent fact-finding missions, however, revealed that the tribal villagers had incurred such state wrath for, firstly, daring to stop paying bribes to forest staff for collecting subsistence forest products in accordance with their customary rights; secondly, for stopping illicit tree felling by powerful interests; and, thirdly, for demanding that the Forest Department itself stop 'scientific' fellings in their forests. *Adivasi Morcha Sangathan* members continued being pressured to join official JFM committees through promises of cases against them being withdrawn, as well as promises of wage work (PUDR, 2001). The incident was a stark demonstration of the hollowness of the devolution rhetoric.

In Uttarakhand, in June 2001, divergence between state and community notions of devolution became starkly evident at a meeting with community representatives, social activists, academics and Forest Department officers for disseminating the findings of the Uttarakhand research study. Community representatives were outraged by the revised Van Panchayat Rules, 2001, which they saw for the first time at the meeting. Asserting that their *van panchayats* had not been gifted to them on a platter but had been obtained through hard struggle by their forefathers, *van panchayat* leaders resolved to launch a popular struggle to demand their withdrawal and to draft alternative rules through broad-based consultation (Sarin, 2001c). The sheer strength of popular anger against the new rules, and the sustained campaign for their withdrawal, led by the *Van Panchayat Sangharh Morcha* over 15 months, has compelled both the Forest Department and the political leadership of the new state to promise their revision.

The research studies in all three states also highlighted the nature of power dynamics within and between communities for control over the forest and major inter- and intra-village equity issues. Poor women, in particular, and the most forest-dependent sub-groups, in general, remain marginalized even in community decision-making and associated federation-building processes. Where such groups have initiated forest protection, they face the threat of being marginalized once the value of the resource has increased through protection. Without effective checks and balances, CFM, on its own, in a context of increasing socio-economic differentiation within communities, is unlikely to provide democratic space for local forest management to the voiceless and disempowered within communities.

## Where to from here?

'Devolution' policies such as JFM thus far represent instrumentalist interventions for obtaining local cooperation in improving forest condition according to traditional Forest Department criteria. This is being attempted on outdated and contested maps of the forest estate. Instead of facilitating holistic forest-lands use and planning, which integrates sustainable livelihoods of forest-based communities, they are extending and consolidating state appropriation of the limited remaining common lands. The 1988 national forest policy redefined the objectives of forest management, which requires re-examining forest categories and forest legislation, as well as institutional structures for forest management developed for colonial interests. Rational re-categorization of the forest estate to acknowledge and prioritize livelihood and conservation needs within a holistic land-use policy framework is a precondition for creating democratic space for local forest management. Without restoring full livelihood and income benefits from local forests to acutely forest-dependent women and men, the sustainability of both the present forest management framework and the 'community' institutions being promoted for its imple-mentation will remain in question.

In a context of rapid change, and the increasing recognition of the multiple values of forests for multiple stakeholder groups, an appropriate framework for sustainable and democratic devolution of forest management can no longer be evolved by forest departments on their own. Holistic forest-sector reform processes with multi-sectoral and multi-stakeholder participation, and multi-disciplinary analysis, need to be initiated. Local people who are heavily dep-endent upon forests for their livelihoods need to be assured a primary voice in such processes through their existing, often long-standing, formal and informal community forestry institutions.

Nurturing democratic, self-governing CFM institutions require a frame-work that ensures tenurial security over community forests, clear boundaries defining communal property rights and empowerment of forest-dependent women and men to make real choices for enhancing sustainable livelihoods in accordance with their own priorities. State interventions need to build upon and democratize existing local initiatives and institutional arrangements instead of seeking to replace them wholesale with standardized state-engineered institutional frameworks such as JFM. Similarly, existing community-based mechanisms for managing inter- and intra-village disputes need to be strength-ened through democratization and by increasing their gender sensitivity, instead of replacing them with inaccessible, ineffective and often corrupt bureaucratic and judicial processes. Inter-village forums or federations of acutely forest-dependent women and men need to be nurtured and empowered for negotiated settlement of disputes and in order to protect their rights and access to forest resources.

Informed participation in decision-making requires improved information dissemination, especially concerning empowering legislation such as PESA, and support for building local capacity to govern and manage forests. Various NGOs and people's movements are already engaged in this work; but addit-ional resources and legal support are needed if forest-dependent women and men are to consolidate and expand on their existing rights in forests. Care must be taken, however, that support does not corrupt the accountability of local leaders and others to the most forest-dependent sections of the population. Experience with 'service-provider' NGOs and elite-dominated 'community-based organizations' suggests that finding ways of delivering support effectively will take innovative thinking. While new mechanisms are developed to deliver support with accountability, pressure should be placed on the state to allow the poorest forest users and their allies to mobilize on their own. Merely eliminating state repression of local forest users will go a long way towards increasing local capacity.

Lastly, the threats posed by market-driven globalization to genuine devolution and democratization need to be addressed. There is an acute need for strengthening global platforms where local perspectives can be shared and appropriate actions identified.

*4*

# Creating Space for Local Forest Management: The Case of the Philippines

*Antonio P Contreras*

## SETTING THE CONTEXT FOR THE CREATION OF SPACE FOR LOCAL FOREST MANAGEMENT

With its formal origins as a state project appearing during the Marcos dictatorship, community forestry began as an instrument to legitimize the state's relationship with rural people, particularly at a time when populations and insurgency were rising in upland areas (Magno, 2001). Although the entry point of the project was to improve forest cover and management, community forestry was also promoted as a project that would improve the well-being of upland forest dwellers. The state, however, aimed to improve well-being primarily to meet its own goals of legitimacy, and as an enticement to participate in state forest projects. This legitimization logic continued to colour policy development during the post-Marcos years as the state forestry agency enlisted communities' help in reforestation and forest protection. By espousing social objectives of poverty alleviation and empowerment, the state gained political legitimacy that allowed it to enter into the domains of villagers' lives and, through them, to increase its control over forest management. The state enlisted communities as a cost-cutting measure, not only to unload itself of transaction costs in protecting the forests, but also to take advantage of voluntarism and participation cultivated among community participants, and to appropriate the social capital of communities as an input for resource conservation.

Despite changes in strategies and instruments (see Box 4.1), what was devolved to communities largely comprised the responsibilities over the protection of, but not the authority to make management decisions about, forest resources (Gauld, 2000). In most cases, community forestry recruited communities as 'participants' in projects, a large part of which were funded by foreign donors and therefore had limited duration, usually five years. The projects fenced people off from the forest by providing them with alternative but rarely sustainable livelihoods, or by providing them with limited access to forest products without giving them adequate support for sustainable markets.

In 1971, the first attempt to address the issue of forest occupancy was launched through the Forest Occupancy Management Programme (FOM). Basically, the programme was designed as an alternative to the earlier relocation strategies adopted. However, its scope was limited only to 'managing' the communities so that they can be removed from their allegedly destructive activities, and not in transferring rights to communities to 'manage' forest resources. Later in the decade, two similar programmes were established by the state, the Communal Tree Farm Programme (CTF) and the Family Approach to Reforestation Programme (FAR), both of which were designed to recruit communities as labour in reforestation activities largely administered directly by the state. All of these programmes had the tacit agenda of stopping the practice of swidden agriculture by making people settle permanently. In 1982, the beginning of a more expanded strategy in community forestry was launched through the Integrated Social Forestry Programme (ISFP), in which communities or individual households were given 25-year leasehold agreements to confine their production systems to communal or household agroforestry farms as a way of providing secure tenure and a source of livelihood. As the leases imposed substantial restrictions on land-use practices, local people were again relegated to roles in which they carried out activities designed primarily by state officials.

In 1991, the Local Government Code transferred responsibility for local forest management to local government units in many locations. Unfortunately, this occurred in the context of a political culture where local political units are rarely interested in forest management unless it offers political or fiscal benefits, especially as the conduit for political and economic patronage, which as often as not led to forest destruction. In fact, where local and progressive forestry officials have already won the trust of local communities, the Local Government Code has compromised the successful implementation of community forestry in some places. Thus, except in rare cases, this type of institutional innovation has failed more than succeeded in bringing forest management closer to communities, despite the stated intentions of the policymakers.

Another development is the Indigenous People's Rights Act (IPRA). IPRA grants extensive tenure rights to indigenous cultural communities over their ancestral lands and resources. A recent Supreme Court decision upheld IPRA, after an extended legal battle. The act was supported by an alliance of the Office of the Solicitor General, acting on behalf of the Department of Environment and Natural Resources (DENR), environmental public interest lawyers and representatives of the indigenous cultural communities. A retired Supreme Court justice questioned its constitutionality, however, invoking the logic that all forest lands are owned by the state – that IPRA, in effect, grants indigenous communities ownership rights and, therefore, that IPRA violates the constitution. The opposition to IPRA highlights the difficulty of breaking free from constitutional norms that limit land-use arrangements to leasehold and

**Box 4.1** *Community-based resource management programmes in the Philippines developed by the state from 1982 to 1995*

Year   Programme

1982   Integrated Social Forestry Program (ISFP)
1984   Rain-fed Resources Development Program (RRDP)
1986   National Forestation Program (NFP)
1989   Community Forestry Program (CFP)
1990   Integrated Rainforest Management Program (IRMP)
       Low-income Upland Communities Project (LIUCP)
       Regional Resource Management Program (RRMP)
       Forest Sector Program (FSP)
1993   Coastal Environmental Program (CEP)
       National Integrated Protected Areas (NIPAS)
       Forest Land Management Program (FLMP)
       Ancestral Domains Management Program (ADMP)
       Integrated Forest Management (IFMA)
1995   Community-based Forest Management (CBFM)

*Source:* adapted from Santos, 2000

exclude ownership of forest lands. It also suggests that attempts to implement the act will meet with opposition and conflict.

This points out another important aspect of the context of devolution in the Philippines today. The transfer of responsibility of managing forests from the state to local communities is not simply a response to fiscal crisis in the state, or an attempt by the state to strengthen its legitimacy on its own terms. It is also the outcome of pressure from civil society (Magno, 2001). Marcos certainly provided support to community forestry to contain the erosion of his government's legitimacy. At the same time, however, the long experience with a dictatorial regime provided the harsh environment to harden-off the seeds of what would soon become an active and vocal civil society. With the departure of the dictatorship, and the re-establishment of constitutional democracy, civil society organizations have provided support for the expansion and deepening of community-based forest management (CBFM), whether by acting as mediators between local forest users and the state, as critics and advocates on behalf of forest users, or both. This, then, provides the silver lining in this rather grim picture of devolution policies in the Philippines. Community efforts to act outside and against state policies, in cooperation with other progressive sectors of civil society, continue to provide insightful critique of, as well as alternatives to, state initiatives. This also is manifested in the emergence of progressive allies in the bureaucracy and in academe that participate directly in the production of policy-relevant knowledge.

These progressive forces face the risk of cooption by the state. For example, community forestry during the post-Marcos era counted on the support of civil society mediators. The enormous financial windfall emanating from tax-free financial rewards for the services of third parties implementing community forestry has generated a whole industry of rent-seeking non-governmental organizations (NGOs). There is also a clash of cultures between cause- and process-oriented NGOs who, in order to take advantage of the new space for local forest management, have volunteered to cooperate with a rigid and output-oriented forestry bureaucracy concerned with projecting success and setting time tables (see Salazar, 1996, and Lynch, 1999, for a discussion). Local initiatives face the threat of co-option, as well, through gradual integration within the community forestry programmes of the state. This threat is difficult to avoid as the land on which these self-initiated efforts operate is legally owned by the state.

It is in this context that devolution policies in the Philippines emerge and are implemented.

## INQUIRING INTO THE CREATION OF SPACE: METHODS, SITES AND DEFINITIONS

The process of data collection and analysis was guided by the principles of participatory research and methodological pluralism. Specifically, the following strategies were used:

1 Research teams in all sites were organized as a tripartite partnership between an academic researcher, a researcher from a mediating third-party institution and a researcher from the local community.
2 Various data-gathering and analysis techniques were used, such as document survey and analysis, analysis of household surveys, content analysis of key-informant interviews, biophysical assessment and participatory analysis through focus group discussion.
3 Community validation of results of analysis was conducted, even as the tripartite structures of the research teams and the participatory nature of data gathering have ensured the visibility of local voices not only in the data but also in its interpretation. Furthermore, some of the partners had data collected from long-term studies.

The 11 sites selected for the study reflect two devolution conditions. Six sites had state-initiated community-based forest management interventions. The remaining five had community-initiated interventions driven by organic processes (in four sites) or third-party civil society-initiated interventions (in one site). As the results of the analysis will reveal, the four self-initiated devolution processes also experience the on-going project of the state to colonize the spaces that have already been opened for local forest management. In addition to the nature of the devolution process, the sites were selected on the basis of the variability in the following variables: tenurial status, forest quality, forest culture, forest economic dependence, level of social capital, and level of organizational capabilities. Table 4.1 summarizes the features of the different sites vis-à-vis these variables. The values of ordinal variables – such as low, medium and high – were determined through discussions among all research team members, comparing current research sites and those from the previous experiences of the researchers.

To facilitate interpretation of the table, the following definitions for certain terms are provided:

- CBFM: Community-Based Forest Management Programme. In this programme, communities are given communal leases by the state in the form of Community-Based Forest Management Agreements (CBFMA), good for 25 years but renewable for another 25 years. The activities involved are agroforestry, reforestation, timber stand improvement of residual forests, and assisted natural regeneration of regenerating stands.
- ISF: Integrated Social Forestry Programme. In this programme, communities are given, by the state, either communal leases in the form of Community Forest Stewardship Agreements (CFSA) or individual household leases in the form of Certificate of Stewardship Contracts (CSC), both of which are good for 25 years but renewable for another 25 years. The forest management activity is limited to the conduct of agroforestry farming with the condition of planting at least 20 per cent of the area to forest trees. Most ISF projects have already been devolved to local government units by virtue of the Local Government Code, while some have been retained by the DENR and are managed as Centres for People Empowerment in the Uplands (CPEU).
- NIPAS: National Integrated Protected Areas Systems. In this programme, communities are used as social 'fences' around protected areas and manage as buffer zones. Tenurial instruments may be granted using either CBFM or ISF models.
- CRMP: Community Resources Management Programme. This programme was later integrated within the larger CBFM programme.
- ADMP: Ancestral Domain Management Programme. This programme is aimed at indigenous cultural communities who are awarded tenurial rights by the state to ancestral lands in the form of Certificate of Ancestral Domain Claims (CADC), good for 25 years but renewable for another 25 years.
- Internal social capital: The web of trust that exists among different groups within the community.
- External social capital: The web of trust that exists between the community and external mediators. This term is used only to differentiate social capital that exists among groups within communities from that which exist between the community and external institutions. Thus, the word 'external' should be taken not as a descriptor of social capital itself – indeed, all social

**Table 4.1** *Summary of features of the sites – Philippines*

| Site | Nature of devolution (programme) | Tenurial arrangement with the state | Forest quality | Cultural dependence on forests | Economic dependence on forests | Internal social capital | External social capital | Organizational capacity |
|---|---|---|---|---|---|---|---|---|
| **State-initiated devolution** | | | | | | | | |
| Quirino | State initiated (ISF devolved) | Individual lease good for 25 years | Degraded | Low | High, but only on forest lands | Weak | Weak | Fair |
| Bicol National Park | State initiated (NIPAS) | None | Fair | Low | Low | Weak | Weak | Weak |
| Camarines Sur | State initiated (CBFM) | Communal lease good for 25 years | Degraded | Low | Low | Weak | Weak | Fair |
| Palawan | State initiated (CBFM) | Communal lease good for 25 years | Fair | Low | Low | Strong | Weak | Fair |
| Cebu | State initiated (ISF–CPEU) | Individual lease good for 25 years | Degraded to fair | Low | High | Strong | Weak | Fair |
| Davao Oriental | State initiated (CBFM) | Communal lease good for 25 years | Fair | High | Low | Strong | Strong | Fair |

| Site | Nature of devolution (programme) | Tenurial arrangement with the state | Forest quality | Cultural dependence on forests | Economic dependence on forests | Internal social capital | External social capital | Organizational capacity |
|---|---|---|---|---|---|---|---|---|
| Civil society-initiated devolution (communities and third-party mediators) | | | | | | | | |
| Nueva Vizcaya | Self-initiated but state legitimized (ISF–CFSA) | Communal lease good for 25 years | Degraded | High | High | Strong | Weak | Strong |
| Laguna | Self-initiated | None | Fair | Low | Low for forest products; high as water source | Strong | Weak | Strong |
| Mindoro Oriental | Self-initiated but state-legitimized (ADMP) | Communal lease good for 25 years | Degraded | High | Low | Strong | Strong | Fair |
| Bukidnon | Civil society initiated but state legitimized | Communal lease good for 25 years | Fair | High but vulnerable | Low | Strong | Weak | Weak |
| Sarangani | Self-initiated but with nascent state presence (through ISF) | Communal lease good for 25 years | Fair | High | High | Strong | Strong | Strong |

capital would be internally possessed by the community – but of the locus of such social capital.

- Strong organizational capacity: High technical capacity and autonomy in decision-making of an organization.
- Fair organizational capacity: High technical capacity, but low autonomy in decision-making of an organization.
- Weak organizational capacity: Low levels of technical capacity and no autonomy in decision-making of an organization.

## IMPACTS OF DEVOLUTION ON THE CREATION OF SPACE

Devolution itself is a contested concept. The statist interpretation that is largely mainstreamed by formal policy should be challenged by the equally powerful but administratively marginalized concept of devolution as an organic process for the self-initiated creation of space. An analysis of the impacts of devolution in the Philippines yields a disturbing yet expected outcome: while devolution processes initiated by the state have opened opportunities for communities to take responsibility for forest management, such space has been bureaucratically constructed to a point that it limits the process of social and political development of communities (Gauld, 2000). Devolution creates space for participation, but not for empowerment. Even in those areas where communities took the initiative of managing their resources, the entry of the state has begun to convert these organically grown forms into bureaucratized structures and processes.

### Impacts on forest quality and livelihoods

The recruitment of communities as partners in forest management has had the purpose, often explicitly stated in policy documents, of protecting existing forest cover from further degradation and denudation, or of taking steps to recover some forest cover and quality. Present in all of the 11 sites studied, the most prevalent forest management strategy was the organization of a local brigade of forest guards, mostly male, to help the state in protecting the forests, both from local illegal users as well as from outsiders. Thus, instead of fostering a developmental process in forest management, a regulatory discourse of the state is extended into the community, and community actors are recruited to augment the policing powers of the state. Community mechanisms are established whereby local policies are formulated to regulate forest use, with such regulation existing in the form of self-inflicted limits on the harvesting of forest products. This makes the task of forest protection easier for the state by effectively exploiting the sense of voluntarism among communities. While volunteer efforts would be much better than coercion, the reliance on unpaid labour can become forms of exploitation when seen against the reality that such volunteer efforts take time away from productive work. In almost all instances, volunteer forest protection work rests on the social capital belonging to local communities and is not in anyway remunerated.

The reward for voluntarism usually comes in the form of the recognition of the status of communities as legitimate residents in a place and the granting of limited tenure. These are tangible benefits clearly valued by participating communities (Lynch and Talbot, 1995; del Castillo and Borlagdan, 1996). Forest cover and quality has also been enhanced in many cases. Too often, however, livelihoods have been compromised in the actual implementation of devolution policies.

In sites in Palawan (Contreras et al, 2003), Camarines Sur (Santos et al, 2003) and Davao Oriental (Morales et al, 2003), all of which are under CBFM, limited harvesting was allowed, and the promise of monetary benefits from resource utilization provided an incentive for forest protection work. However, when the DENR suspended all harvesting permits of communities, communities were stripped of these limited tenure rights and the incentives they provided for good forest management. There was at least one documented case in this study that is also true in other sites not included in the CBFM site in Camarines Sur, wherein illegal forest extraction resumed after the suspension of permits with the tacit knowledge of the local forest protection officers. This case highlights the unsustainability of a forest protection strategy that rests on a state-initiated appropriation of 'voluntary' labour where economic benefits are limited.

Spaces for local forest management also are created through state support for alternative livelihoods, only to be eroded by the absence of adequate development of credit and markets. From among the 11 sites studied, only the site in Mindoro Oriental (Paunlagui et al, 2003) had a well-defined market and credit-support mechanism, which was not even provided by the state but through the mediation of a church-based NGO. In addition, two other sites in Quirino (Dizon et al, 2003) and Bukidnon (Suminguit et al, 2003) had access to credit but had weak links to markets. It is also interesting to note that in the three CBFM which had limited utilization privileges (Palawan, Camarines Sur and Davao Oriental), access to markets and to credit are also, at best, tenuous. In all of these sites, the community is forced to deal with forest product markets controlled by a single or a few buyers who can dictate the price.

Another strategy by DENR programmes used to arrest the erosion of resource quality is through the adoption of agroforestry conservation technologies and soil conservation measures. This strategy again intersects with livelihoods in the sense that aside from providing an avenue to divert the farmers from destructive upland farming practices and to encourage otherwise shifting cultivators to become sedentary, it also creates an opportunity for earning income. Hence, agroforestry is basically the domain for which positive income benefits can be realized. However, the absence of strong support systems, coupled with the reality that most of these lands are of marginal productivity, has compromised the full realization of such economic opportunity. While agroforestry was promoted as a land management strategy in 9 of the 11 sites, it is more often the case that forest-based farming is not the major source of livelihoods. In fact, in two sites, off-farm income sources are higher than on-farm income sources. For example, in the site in Davao Oriental, 55 per cent of the income of community members came from off-farm sources and only 35 per cent from agricultural harvest. In the site in Mindoro Oriental, which is primarily composed of indigenous peoples, income from forests and forest lands constitute only about 31 per cent, while income from private employment alone constitutes 48 per cent. This indicates the fact that agro-forestry may no longer be a viable domain for income creation in some upland areas.

Furthermore, agricultural production systems are not even enough to provide food security to some communities. In the site in Camarines Sur, it has been noted that sales from agroforestry farms are not even enough to feed the family. In the absence of off-farm employment opportunities and stable product markets, the introduction of agroforestry may again not provide a viable income source. The story here is similar to forest products, where the absence of stable market linkages, compounded by the perishable nature of crops and relatively distant and poor-quality farm-to-market roads, makes the farmer totally vulnerable to middle persons who control the market. In fact, only two sites (Quirino and Nueva Vizcaya) (Torres et al, 2003) showed evidence of the existence of efforts to provide adequate ground transport support through the maintenance of farm-to-market roads, and, in the case of Quirino, bridge access to farmers.

Reforestation and plantation development strategies are also deployed to improve forest cover. In fact, reforestation as a strategy was practiced in seven sites. This approach provides only short-term sources of employment. In sites studied in Palawan, Camarines Sur and Davao Oriental, all CBFM sites, as well as in Cebu (Chiong-Javier et al, 2003), Quirino and the Bicol National Park (Pulhin et al, 2003), reforestation activities were seen in terms of employment opportunities. It is also in this activity that women's participation in resource management, seen in the preparation and care of planting stock, was partic-ularly visible. The problem with this particular land development strategy is the shortness of the employment opportunity, coupled with the prohibition to harvest the planted tree in the future. As a result, this activity is an unsust-ainable source of income. Furthermore, the failure to provide rights to harvest trees planted, even in those areas that have approved utilization rights for non-timber forest products (NTFPs), has denied some communities the incentives to maintain the crop after they are planted. In fact, in Mindoro Oriental, the Mangyan community refused to engage in reforestation. They feared that the project would not bring economic benefits to them, and that the presence of reforestation activities would only give the state reason to further intervene into their affairs. They perceived that the planted crop would not be theirs and the presence of trees planted through state financial support might reduce their stake in their ancestral domain.

The most visible benefit in at least eight of the study sites is the awarding of tenurial instruments. All of these, however, are in the form of leasehold agreements with 25-year limits, renewable for another 25 years. In practice, then, indigenous cultural communities in Mindoro Oriental, Bukidnon and Sarangani (Duhaylungsad et al, 2003), all of which have claims on ancestral domains prior to the existence of the Philippine state, have to recognize the latter's ownership of the land by 'applying' for such tenurial instruments and allowing themselves to be subjected to state regulation. The state, acting as a landlord, defines for its 'tenants' (the local communities) the terms for their occupancy. Communities do have input in the preparation of forest management and resource utilization plans, but only according to specifications issued by DENR. These specifications extend beyond complying with a fixed format, but even go to cover the management decisions, such as choice of species and the harvesting of secondary forest products. This leaves very little room for local communities to influence resource management and policy. They can refuse to comply or adopt measures that will create coping mechanisms to subvert or 'go around' the terms of the policy. But this courts the risk of state reprisal. We have already mentioned the controversial cancellation, at a national scale, of resource-use permits. This was a result of a violation of the terms of the permit granted by the state *in just one site* (not included in this study) in Mindanao. Tenurial instruments such as stewardship contracts and community management agreements are not only overly restrictive, but could also face cancellation in the event that violations anywhere are detected by the state.

## Impacts on welfare and social development

The failure of formal devolution policies to provide adequate livelihood opportunities is compounded by their failure to address social development concerns, such as health and education. Impacts on welfare occurred most noticeably where communities were organized. The importance of social development is highlighted by the case from Laguna (Tolentino et al, 2003). Here, the local community has no tenurial rights over the watershed. In fact, a big part of the forest is privately claimed. But this did not stop the community from organizing to protect the watershed that provides the source for their domestic water needs. The key to their success was that they had a fair amount of social development infrastructure already available. Their primary economic interest in the forest – its ecological service of providing adequate water – was thus consistent with the DENR's agenda of keeping the forest intact. They also had the skills and resources to organize forest protection. The state provided the community its social infrastructure needs, such as schools, roads, health facilities, electricity and access to transport, that have provided it with the social development space needed to effectively manage forests.

This is, however, not the case in most upland sites where communities are socially marginal: their incomes are below the poverty line, and they have poor access to adequate educational, health and other social development services. The site in the Bicol National Park is a good case to contrast with the site in Laguna. The former captures the typical upland condition of social, economic and political marginalization. Here, efforts of the state and its contracted intermediaries are all geared towards the establishment of local capacity to manage (that is, to protect) the forest resource, even though adequate social development services are not provided.

Only two out of the other ten sites reported some attention given to social support systems within the context of the forest devolution intervention. The site in Quirino reported some access to health services, such as a clinic and potable water supply through the establishment of a water impounding system. In Mindoro Oriental, infrastructure services were established in connection with the earlier state devolution project, the Low-Income Upland Community Programme (LIUCP). In other sites, the provision of social development services are relegated to the local government units, which, in some instances, are seen as a competing interest. In Palawan and in Nueva Vizcaya, the local government units have shown a predisposition for looking at the forestry people's organization as a rival for political loyalty. This attitude has prevented people's organizations from taking up more active social development interventions for fear of being accused of usurping the jurisdiction of local governments. In the final analysis, it is only when the Barangay leadership and the forest people's organization have good relations that cooperation between the two local institutions can be mobilized to provide the necessary social development projects, as in the case of the sites in Mindoro Oriental, Quirino and Sarangani.

In general, where benefits have accrued to local villagers, they have accrued to well-educated, wealthier individuals (Rebugio, 1996). There is, however, one domain where devolution policies, as externally mediated processes, may have brought social development into some communities. This is in promoting gender equality. The evidence in at least three sites (Nueva Vizcaya, Sarangani and Davao Oriental) shows that the presence of a gender-sensitive mediator can facilitate the mainstreaming of gender issues. However, even in these cases, the involvement of women is limited to short-term employment in nursery operations and livelihood activities. There is no active agenda to fully mainstream women's participation in key decision-making, as well as to address social development issues that are specific to women. Thus, even in cases where gender is recognized as a development concern, the gender and development (GAD) agenda of promoting gender equality through women's empowerment is still not fully achieved. It is, then, safe to expect that the situation is worse in those sites where there is a total failure to recognize gender and women's concerns. In 8 of the 11 sites studied, which include some of those that are self-initiated, such as Mindoro Oriental and Laguna, there is a total absence of a women's agenda. Worse, there are reported cases of women's exclusion from benefits, such as in sites in Cebu, Camarines Sur and Quirino.

It is understandable that the focus of forest devolution should centre on forest management and protection issues. The effectiveness of any devolution strategy, however, is largely enhanced by the presence of social development, as indicated by the presence of sustainable support services such as schools, health centres, infrastructure support and other facilities. As dramatized by the case of the Bicol National Park, no amount of institutional building can provide long-term sustainability for resource management in the absence of secure, developed and healthy communities. Unfortunately, the devolution model adopted by the Philippine state vis-à-vis forest management has failed to deliver this kind of development.

## Impacts on decision-making

The agenda of the state in forest devolution – to consider the community and its social capital as inputs to forest management and protection – takes the form of organizational structures and processes to recruit communities as 'partners' (unpaid volunteers). In Bicol National Park, the level of state intervention was remarkably intense. After being forcibly relocated from within the protected area, the community was subjected to a heavy dose of institutional interventions. The density of state and third-party mediators contracted by the state to organize and prepare the community for forest management work – numbering nine organizations, including DENR – led to an amazing situation where in just one village, there emerged seven people's organizations. In principle, a representative policy-making body, the Protected Area Management Board (PAMB), established in accordance with NIPAS, provides the opportunity for these people's organizations, who have reserved seats in the board, to articulate their voice with representatives of local government units, civil society, and other state agencies who are also members. However, this fixation on building organizations has neglected social development, even as formal policy has yet to provide adequate tenurial rights to people in accordance with the buffer zone model of protected area management. Thus, even as patches of reforestation and a decline in illegal forest poaching are happily reported by the state, as indicated by a decline in the number of incidents reported from 11 in 1996 to only 1 in 1999, the community remains vulnerable. And their participation in the PAMB remains a symbolic gesture to mechanically allocate space for their presence, but not for their meaningful engagement. This is because while you have a very dense organizational setting, there is an utter failure to provide adequate human services that are required to ensure social security seen in secure tenurial rights, adequate livelihood and access to respectable social development services.

The rigid and bureaucratic structures and processes that accompany formal state devolution threaten those sites where organic devolution has occurred. In the site in Nueva Vizcaya, which has seen the full effects of the transition from self-initiated to state-mediated devolution, the autonomy in decision-making about forest resource management has already shown signs of being eroded, with some decisions being gradually taken over by state-imposed parameters, such as choice of species and management systems. A similar trend is feared in the site in Sarangani. The local community is wary of engagement with the state through the ISFP, as the programme would greatly diminish local control over the resource in exchange for 25-year leases. Communities participating in social forestry programmes are required to conform with certain management practices, including the prescription of certain exotic species for planting. Local leaders are hoping for a change in policy, which might prove to be difficult to achieve considering that such policies are national in scope and would require a national-level effort. The site in Laguna, which is largely still an NGO site, has already seen the ill effects of the Local Government Code, legislation that is designed to devolve resource management to local political units (see Box 4.2). There, the local government unit has begun to interfere with the manner in which the watershed is being managed.

---

**Box 4.2** *Excerpt from the case study of Balian, Pangil, Laguna*

While the community has shown its tenacity to manage whatever problems it confronts, it seems that such capacity is brought into test again in recent years. Interestingly, this occurs at a time when the local government units are given the power to raise revenues from its natural resources. While the management and supply of water in this village was left entirely in the hands of SPBTI [*Samahan ng Balian para sa Pagpapauwi ng Tubig Inumin:* the local people's organization], the Barangay Council is beginnning to show its interest in this particular resource. At first, it negotiated development assistance from Japan to construct a water reservoir in the same place where SPBTI's reservoir is located. It also provided for the installation of new water pipes that would supply water to the village. All of these were done without the participation of SPBTI. The Barangay Council reckoned that since the Local Government Code of 1991, otherwise known as Devolution Law, provides it with power to develop the natural resources within its territorial jurisdiction, then it could do what it wanted to with those resources. However, when the Barangay Council wanted to take control of the water resource, then conflict with SBPTI became apparent. This conflict was even manifested in the last election of officers of SBPTI. The Barangay Council wanted to field the whole Barangay Council as a candidate for the President of SBPTI.

*Source:* Tolentino, Plopino and Jacinto (2003, p155)

---

The formation of organizations can also have exclusionary impacts within communities. These often reflect existing networks and power structures. One salient feature of community forestry in all of its variations, as applied in the Philippines, is the formation of forest people's organizations. These organizations are composed of individual households who are either beneficiaries of individual leasehold agreements, as in the case of ISFP, or have formally joined the group to benefit from devolution. This strategy necessarily excludes those members of the community who are not formally affiliated with the organization. Except for being considered as forest occupants, or as families actually cultivating forest lands, there are no rigid exclusionary rules. However, despite this, local power relations along clans, as well as kinship rivalries and loyalties, and geographical distance from the centre of the village, could serve as barriers to membership by some community members. This leads to a situation where not all forest users and occupants are members. In the site in Cebu, for example, only 69 per cent of the households are members of the people's organization; in Davao Oriental, only 60 per cent are members. In Bicol National Park, it was even only 27 per cent. To some, a figure above 50 per cent may be high. However, the issue here is not that most people are involved, but that some people are excluded and therefore are either denied the benefits of the project, or are operating outside the collective norms established by the organization and are beyond its reach.

The exclusion of other members in the community may also lead to hostility among groups. In Camarines Sur, those who were excluded from the people's organization have instigated a smear campaign against the latter with the local government units by accusing them of illegal logging. This eroded the credibility of the organization with the local government unit. In Nueva Vizcaya, state support for the forest people's organization through the ISFP project was even used as an excuse by local government officials to divert social development activities away from them. Local officials instead favoured those who were not members of the people's organization, and were therefore not participants and beneficiaries of the project, allegedly to promote equity within the village. It is interesting how an intervention by one government agency, aimed at fostering equity, could lead to a situation where equity arguments are used by another agency to deny development assistance.

It is natural for organizations to emerge at the local level, particularly when some collective action is required to achieve common goals. In the four sites where self-initiated mobilization towards forest management occurred (Nueva Vizcaya, Mindoro Oriental, Sarangani and Laguna), the presence of an environmental crisis has motivated local communities to act. What motivated communities was the realization that resource degradation could compromise their livelihoods. In these cases, the process of organization building was firmly based on organic foundations. The site in Mindoro Oriental, however, was unique. Even as the community acquired the capacity to resist and influence policy by refusing to conduct reforestation activities, there was a palpable dependence on a third-party mediator, a church-based NGO, in running its affairs. A dependency relationship emerged between the mediator and the community, raising doubts as to whether the community could act on its own in the event that the NGO withdrew, as it had already begun to do.

It is clear from these discussions that even as opportunities for technical capacity-building are provided by formal devolution, the structure of the policy limits the translation of such technical capacity into full autonomy in decision-making. The state has a tendency to create bureaucratic structural limits on an otherwise participatory development ideology. This erodes the organizational capacity and influence of even those institutions established through organic collective action.

## Conditions that Influence the Creation of Space

The development of the devolution policies of the state in the Philippines, as earlier mentioned, has followed a uniform trajectory. The state attempts to harness community resources and to co-opt community forest management, progressively transferring resource management responsibilities without full devolution of decision-making authority. Communities, however, resist this form of devolution. The outcome of struggles over forest management has depended upon the level of social capital, the degree of cultural and economic dependence of the community on their forests, and the presence of the state and external institutions.

### Social capital

In the 11 sites studied in the Philippines, the most pervasive enabling factor is the presence of a strong social capital, both within communities and externally as communities relate with mediators. Social capital here refers to the collective capacity to act together based on reciprocity, norms and networks. In fact, there is evidence to show that strong social capital, both internally and externally, is present in all three sites that exhibited the capacity to perform community-level policy-making (Mindoro Oriental, Davao Oriental and Sarangani). Furthermore, even where external social capital was weak, but internal social capital was strong, community-level policy-making was also present, as in the two cases that were products of community-initiated devolution (Nueva Vizcaya and Laguna). In contrast, all of the sites that have low internal as well as external social capital (Bicol National Park, Camarines Sur and Quirino), while possessing some levels of technical capacity, had no autonomy in decision-making.

Another factor that contributed positively to a community's level of influence is a history of self-initiated devolution. In fact, even as it seems too obvious, it is still useful to point out that all sites that have histories of self-initiated action are also the sites with high levels of social capital. The combined power of high social capital and a history of self-initiated collective action effectively deflected the erosive powers of state devolution policies. Formal devolution policies were seen as threats to local community action in Laguna and Nueva Vizcaya. However, in both sites, the presence of strong internal social capital and a history of self-initiated collective action were used by the communities to reinforce their own autonomy in decision-making. In contrast, Davao Oriental is the only site that was exposed to high formal policy presence through the CBFM and for which the devolution strategy was state initiated. In this site, the community possessed high levels of trust among its members, coupled with a healthy relationship with its external mediating institutions. The external NGO contracted by the state to socially prepare the community for full transfer of management responsibility, in accordance with the structure

of the CBFM policy, had been very effective. Nevertheless, while the organization possessed technical resource management capacity, it did not possess
adequate levels of autonomy in decision-making. These examples show that
organically grown social capital and a history of self-initiated collective action
possess more power to deflect threats from the state than when social capital
is built through the activities of external mediators.

## Cultural and economic dependence on forests

A high level of cultural and economic dependence on the forest was also seen
as a relevant causal factor in influencing devolution policy outcomes. The
capacity for community-level decision-making, in particular, seems to be
influenced by a strong forest culture. Except for the site in Laguna, which has
a weak forest culture and low dependence on forests, all sites with strong
organizational capacity also have strong forest cultures (Sarangani and Nueva
Vizcaya). High cultural dependence seems to be associated with sites that
experienced civil society-initiated devolution. In fact, of the five sites where
devolution was initiated by communities and third-party mediators, instead
of by the state, only the site in Laguna has exhibited low cultural dependence
on forests.

It is also important to point out that economic dependence on forests is not
directly related to forest culture. In fact, there are sites where forest culture is
low but economic dependence is high, even as there are also sites where forest
culture is high while economic dependence is low. What is interesting is that
the first case (weak forest culture but high economic dependence) seemed to
have been found in sites that experienced state-initiated devolution; the second
case (strong forest culture but low economic dependence) was seen in sites that
experienced non-state-initiated devolution. However, in the latter cases,
organizational capacity is not strong.

The community in Laguna is interesting in this regard. It does not possess
a 'forest' culture in the sense of having a close connection with the forest. Its
apparent dependence on an ecological service, watershed protection and a
worldview that values ecological services, generally, has enabled the community to take stewardship of the resource. This is extremely significant since
it points to an interesting finding – that an interest in ecological services can
be the basis of collective action. The Laguna site can also be used to deconstruct
'culture'. Forest culture is something beyond the classical anthropological
construct of rituals and symbols associated with forest, and can include urban-
based perceptions. However, the case in Laguna also dramatizes the elitist
nature of this discourse. The community in the site has the luxury to manage
the watershed in an exclusionary way because it has satisfied some of its basic
needs. One can only ponder how different the collective response could have
been had the community's main livelihood rested on activities within the
watershed, circumstances faced by a significant majority of forest-dependent
and forest-based communities. One can also imagine how the lowland community's strategy could have differed (and the level of conflict that could have

emerged) had the watershed in question not been a mere 31-hectare area without defined settlements, but a large tract within which forest-based communities pursued their livelihoods.

## Presence of state and external institutions

In the short run, a strong state presence enhances the focus on physical activities that are geared towards forest protection, as well as the provision of some limited livelihood alternatives. The site in Cebu, which remains a DENR-managed Center for People Empowerment in the Uplands (CPEU), clearly demonstrates the favourable effects of direct state connection. But, as discussed earlier, the sustainability of these remains questionable. The community in Cebu, despite its strong internal social capital, has very few and weak linkages with other external actors, with the community depending mainly on DENR as the mediator.

External actors and institutions, however, also behave in a complex manner. For example, the presence of a high density of external institutions has delivered some social services in sites in Mindoro Oriental and Davao Oriental. However, similar actors spelled disastrous effects in the Bicol National Park, where there were simply too many cooks. Furthermore, there is fear that too much mediation can foster a culture of dependence, as in Mindoro Oriental. In fact, as pointed out earlier, the effectiveness of a mediator in constructing social capital within communities may not necessarily lead to the development of autonomous decision-making capacities, as dramatized by the case in Davao Oriental.

On a different issue, the presence of institutions that mediate between the state and the community may also lead to the erosion of the mediator's credibility. In the site in Camarines Sur, the honest efforts of the mediating NGO were undermined in the eyes of the community by the bureaucratic limits imposed by the DENR. Unrealistic reforestation targets and delays in fund releases courted the ire of the affected community. Even though the DENR was actually the source of the problem, it was insulated by the presence of the mediating institution that absorbed the hostility of the community.

Relationships with units have also been mixed, with some providing support – as in the case of the site in Sarangani – and others that are less friendly and that consider the forest people's organization as a competing interest, such as the sites in Palawan, Camarines Sur and Nueva Vizcaya. Others are also blatantly hostile, such as the site in Laguna (see Chiong-Javier, 1996, for a general discussion). As pointed out earlier, local government units are recipients of devolved authority for small-scale forest operations, including community-based forest management. Almost 98 per cent of the ISFP sites previously held by DENR have been devolved to the local government units, one of which is Quirino. As reported in the site, the latter did not sustain the initial efforts of the DENR in providing support to the community prior to devolution to the local government unit. Furthermore, it is reported that the local government unit is exacting land development taxes on forest-land users on their activities within lands covered by the tenurial instrument awarded by the state. This has the effect of providing a disincentive to improve land since doing so would lead to the payment of higher taxes. It is also a reason to illegally open more areas in the forest zones where additional land development will not be subjected to tax, since these will fall outside of the authorized area of operation covered by the tenurial instrument. Appeals to the DENR have led the state to issue a statement to the effect that areas covered by stewardship agreements are exempted from real property tax since they do not fall under the category of taxable forest lands. The community continues to pay taxes to the municipal assessor's office, however, for fear of losing their use rights over the land. Furthermore, the local government is subverting the policy by saying that the taxes they are collecting are not property taxes, but land development taxes. There are local government units that are, indeed, supportive of community forestry, as illustrated by the good reviews that the governor of the neighbouring province of Nueva Vizcaya has received for his honest efforts to foster community-based forest management. The case in Quirino illustrates only one of the instances in which this model of devolution has failed.

In the site in Sarangani, there is a complete juxtaposition between and among the *barangay* leadership, the organizational structure for local forest management and the traditional structure of leadership. In this case, the local government unit, which also reflected the traditional leadership structures of the indigenous community, was also the mediator for the entry of state devolution policies. As correctly pointed out in the case report for this site, 'the devolution process occurred not because of the space created by state policies to transfer forest authority to local communities. It is a local historical conjuncture that coincided with the legislation of laws and administrative orders' (Duhaylungsod et al, 2003). What enabled the local government unit to be an effective mediator for the entry of the state in local forest management – while at the same time providing safety nets to protect the community from the disarticulating elements of state intrusion that are now beginning to be manifested – is the fact that it was not considered as an 'other' in relation to the structure for local forest management. Furthermore, the presence of indigenous practices for fostering social cohesion, as well as the high level of trust that the community has in its local community leader, has provided substantial social capital within the community (see Box 4.3).

Special attention should be devoted to the presence of mediating bodies that hope to bring together diverse and often competing interests. As an avenue for policy-making and conflict resolution, these bodies offer enormous opportunities for creativity. Unfortunately, the experience in two of these bodies – namely, the Protected Area Management Board (PAMB) in Bicol National Park and the Palawan Council for Sustainable Development (PCSD) in Palawan – leaves much to be desired. Box 4.4 presents an excerpt from the case study conducted in the Bicol National Park, while Box 4.5 presents an excerpt from the study conducted in Palawan. In both cases, the existence of multipartite bodies fails to accommodate the interests of communities; these bodies

---

**Box 4.3** *Excerpt from the case study conducted in Upo, Maitum, Sarangani*

The influence and political capacity within the community centres around Kubli, with the collective support of his Barangay Council. Kubli, together with his tribal chieftain (*datu*) brother, is able to generate support and cooperation among his T'boli constituents because of the strong overlay of the traditional social capital of reciprocity and the *datu* social structure of T'boli society. In particular, *tenggawa* , the T'boli tradition of work cooperation, and the ethic of *s'basa* (reciprocity principle), has been fully utilized in the work brigade for the various infrastructure projects (tree planting, road improvement, water development, Women in Development of Sarangani (WINDS), and the forest brigade). The *kasfala,* a traditional system of conflict resolution that is headed by the *datu*, is still a powerful instrument. Furthermore, Kubli's constituents regard him with respect and trust. He is perceived to have a leadership character of transparency, dedication and commitment. In recognition of his impressive community leadership achievements, he received leadership awards from the municipal government in the 1999 celebration of the town fiesta. There is assurance that Kubli's leadership initiative will be continued even after his term as *barangay* captain. One of his council members, Edwin Ganuan, has been identified by the community as having leadership potential. It is also worth noting that out of the 11 members of the Barangay Council, there are three women.

*Source:* adapted from Duhaylungsod, Buhisan and Duhaylungsod, with assistance from Kusin and Duhaylungsod (2003, p202)

---

themselves possess distinct interests that compete with those of the community. In both cases, it is apparent that the dominant ideology and worldview that is reflected in these bodies, in addition to other organizational factors, tends to limit instead of enhance the attainment of the local community's interests. Furthermore, as indicated by the experience in Palawan, NGOs may also possess an agenda that would be detrimental to the promotion of community forest management.

In summary, it can be said that space for local forest management was largest where communities were able to resist or localize state efforts. This capacity was strengthened by the presence of a strong social capital, further supported by external facilitators as well as by local institutions. A strong forest culture also enhanced the creation of space. In the end, self-initiated devolution processes are seen to be more conducive for local forest management than state-initiated ones.

**Box 4.4** *Excerpt from the case study of Bicol National Park*

The Protected Area Management Board (PAMB)composition is just a carry-over of the former [Bicol National Park] Core Group; hence, it represents a strong societal interest towards park protection instead of meeting the economic needs of the immediate community. Even three of the four non-governmental organizations (NGOs) in the PAMB cannot fully represent the sentiments of the local people since they were also part of the relocation/demolition project. In essence, the present PAMB composition restricts the very space that the National Integrated Protected Areas Systems (NIPAS) law has created for local forest management.

With the PAMB's inherent bias in favour of the broader societal interest, local people are simply seen as a means of achieving the objectives of forest rehabilitation and protection. Hence, people have to stay away from the periphery of the park to contain their potentially destructive activities. While this strategy seems to work at present, the situation has posed a limit on the potential contribution of the local people in sustainable park management. The fact that there were incidences of intentional burnings seems to indicate that the present situation may not be the appropriate arrangement as far as the management of the park is concerned.

*Source:* adapted from Pulhin and Pesimo-Gata, with the Ramilo (2003, p63)

**Box 4.5** *Excerpt from the case study of the CBFM area in* barangays *San Rafael, Concepcion and Tanabag, Puerto Princesa City, Palawan*

The Palawan Council for Sustainable Development (PCSD) was established, making the situation in the island unique. It is only here where a council, which is a multi-sectoral body, exists with the task of acting as a policy-making body with oversight functions over all environment and natural resource concerns in the island. This council was a result of civil society mobilization that demanded the creation of a local mechanism that would safeguard the environment of Palawan, which was touted to be the last frontier of the Philippines. The Department of Environment and Natural Resources (DENR) is represented in the council through its secretary, which also includes as members the representatives of the local government units (LGUs) and local civil society organizations. Despite its multi-sectoral nature, and as an offshoot of its historical formation, the PCSD is dominated by civil society actors, most of which bear a protectionist environmental agenda. Community-based management of resources, while technically under the coordination of the DENR, still has to pass through the scrutiny of a PCSD that has shown a predisposition in favour of local government units managing common forests and against the direct involvement of people's organizations.

*Source*: adapted from Contreras (2003, p87)

## CONCLUSIONS

The following conclusions can be derived from the analysis of the 11 sites:

1   The positive impacts of devolution policies were seen in the form of gains in visibility and legitimacy of communities in the eyes of the state.
2   However, losses have occurred in terms of the negative and restrictive impacts on communities of the state's vision of what forest management is, the types of economic options available and the bureaucratization of community-based institutions.
3   Community organizing activities that are part of state devolution policies only generate superficial organizational capacities and alliances due to the fact that they are not organically based on natural processes, as well as their reliance on bureaucratic parameters.
4   Third-party mediators/facilitators provide support, even though they are sources of problems, particularly when their agenda clashes with a community's vision of forest management.
5   Self-initiated devolution provides more effective venues for local forest management compared to state-initiated devolution, even as it faces the threat of increasing integration within state projects that have the effect of weakening community power.

Despite the general failure of state-initiated devolution policies, spaces still exist that allow us to imagine and offer alternative models. These models are necessarily informed by a different worldview, but are empirically grounded in the interface between state-initiated interventions and organically grown local action. As a major finding of this study, social capital emerged as the most consistent explanatory variable for increasing a community's influence over decision-making, not only about forest resource management, but also about livelihoods, well-being and policy. A high level of trust and the presence of local alliances, networks and institutions help foster the creation and maintenance of local capacity to represent a community's interests, resist external pressures and localize external processes. Social capital is also important in influencing the capacity of the community to handle the state, which, no matter how problematic, has to be dealt with and fortunately is not a static institution that escapes the possibility of reform. It is in this context that social capital becomes central in developing alternative local forest management arrangements.

In the Philippines it is difficult to recommend a total transfer of authority to communities as this will require social and constitutional change, not to mention possible environmental consequences. What can, at least, be done is to support processes that politically construct social capital, both among forest-based and forest-dependent communities, and across society and among different competing interests. Institutional arrangements are required that address not only the need for more direct control by communities of their resources, but also multi-stakeholder situations with conflicting interests. A society with strong social capital can have the capacity to recuperate from the destructive effects of the state. It can also construct alternatives that may, in the end, lead to the reform of the state. This requires, however, that that society values ecological and political health and is provided with democratic spaces.

The focus, therefore, of advocacy should no longer be on developing alternative techniques for better forest management, but on developing alternative ideologies and worldviews and the institutions to support these alternatives. It is no longer enough to advocate for devolution in forest management. The challenge is to advocate for deconstruction of the dominant forest ideologies and practices at both the level of the state and the community.

In the Philippines, three innovations have emerged in the context of implementing state devolution policies that can be offered to address these concerns, all of which have been formed with active civil society advocacy. The implementation of these innovations will also provide opportunities for civil society participation. These are:

1   the existence of a PAMB, with representation from civil society and from local communities, as a multi-sectoral policy body to govern protected areas;
2   the unique case of the PCSD as a provincial policy body with direct connections to national-level institutions, even as it provides a venue for representation from local institutions, including local civil society actors and communities, on matters relating to environment and sustainable development; and
3   the Indigenous People's Rights Act.

The first two models address the concern for multi-stakeholder situations, while the third addresses the concern for direct community control over resources. Despite their problems, these three innovations, at least in theory, give more direct control over resources to local communities and provide a venue for reconciling competing interests. The reality of a PAMB that does not provide adequate voice to the communities in the Bicol National Park, and of a PCSD that is a tool for local political agenda, only dramatizes the risks that alternative structures may present. But there is reason to be hopeful. As of now, the structural space provided for NGO and local community representation in both the PAMB and the PCSD is promising. The presence of a healthy and pluralist civil society can provide safety nets and countervailing forces to check not only the excesses of civil society, but also the threats from the state. This was highlighted in the small victories of local communities reported in sites in Sarangani and Laguna, with their organically grown capacities to challenge the state, both at local and national levels.

# Whose Devolution is it Anyway? Divergent Constructs, Interests and Capacities between the Poorest Forest Users and States

*David Edmunds and Eva Wollenberg*

As indicated in the previous chapters, devolution policies have often done little to help local forest users shape what forests look like, protect and promote their livelihoods, or establish and maintain their own decision-making processes. In this chapter, we assess the outcomes of devolution policies across the three case study countries and offer an explanation for why these outcomes occurred. We believe the explanation lies in divergent constructs, interests and capacities between the poorest forest users, on the one side, and foresters and other public officials on the other.[1]

These constructs are a set of habits of thinking and acting, supported by institutional rewards and sanctions, that are difficult to break from or argue against, similar to Said's (1978) explanation of how orientalist discourse works. The rewards and sanctions create specific incentives for foresters and forest users, which might be income-earning opportunities or the respect of colleagues and neighbours. Because public officials have greater financial resources, media access, legal knowledge and other sources of power, they have the capacity to act on their constructs and interests at the expense of the poorest forest users.[2]

Our findings show that the state's constructs and interests have remained dominant despite the increasing role of diverse forest users, including even the most disadvantaged rural people (Grimble et al, 1995; Mayers and Bass, 1999; Anderson, et al, 1999). Forest departments in all three countries were explicit in statements and policy documents about promoting forest management that emphasized timber production,[3] watershed protection, fire suppression, soil conservation, enhancement of environmental values or maintenance of forest cover. These goals are, of course, not necessarily incompatible with those of forest users (Salafsky et al, 1999; Tiffen and Zadek, 1998). Unfortunately, the goals have been pursued even when they conflicted with local livelihoods, cultural values and management systems.[4] Where there was a shared rhetoric of local empowerment and sustainable livelihoods, the implementation of devolution policies was still characterized by the misidentification, misrepresentation and exclusion of other groups' interests, particularly local forest users, which ultimately has contributed to less equitable outcomes than could have been achieved (Dubois, 1998; Ramírez, 2001).[5] Our studies prompt us to ask whether it is now time to give more attention to seeking a more equitable balance of interests in forest management.

In the rest of this chapter, we examine in detail how state constructs and interests diverged from those of local forest users. We look first at divergences in the outcomes of local devolution policies in terms of what forests should look like, who should benefit from them and how, and who should have the right to make decisions about them. We then examine the key state strategies underlying the divergence – namely, how the state used taxes and regulations, contracts and the manipulation of local organizations as mechanisms to maintain decision-making control over forests and forest products even after devolving sometimes significant *de jure* control of forests to local users.

## DIFFERENT VISIONS OF FOREST QUALITY

In the three countries studied, devolution policies and the policy-makers and bureaucrats who interpret them treated forests as entities to be managed primarily for timber, commercial non-timber forest products (NTFPs) or environmental services. According to the policies and their implementers, local livelihood needs can and should be integrated with these uses; but the state's priorities should predominate. This is clearly reflected in the state requirements for forest protection, restrictions on forest use, and continued reliance of forest departments and donors on forest cover and forest stock as the primary indicators of the success of devolution policies. By this measure, devolution policies have succeeded. Forest cover and forest stock improved in nearly all of the case sites, either through regeneration, as in much of India, or planting, as in much of the Philippines and China. Then, why weren't many of the local forest users we spoke with happier?

One major reason was that the state focused on certain forest uses to the near exclusion of all others. Whereas forestry officials sought to maximize timber production (and the revenues obtained from timber taxes, etc),[6] local forest users more often allowed some cutting of these timber species for poles, allowed useful competitors to grow in some areas of the forest, and even discouraged timber species where competing fodder species were important (Poffenberger, 1990; Sarin, 1998). Similarly, local forest users valued ecological services, but balanced interests in watershed protection and biodiversity with needs for resource extraction (Western et al, 1994; Borrini-Feyerabend, 1997). In our study sites, where local people managed forests at their own initiative, they sought to accommodate these multiple objectives. Devolution policies' narrow focus on one or two dominant objectives for forest management was inconsistent with multiple livelihood objectives of local users (Campbell 1992).

The gap in visions was starkest in India and the Philippines, where timber and commercial non-timber forest product species dominated reforestation and afforestation efforts at the expense of species used for medicines, wild foods, fodder, local construction, arts and even the protection of soils and water sources. Locals were also asked to establish plantations on land used for grazing or as agricultural reserve areas. At the same time, the most valuable forests were protected to meet conservation goals, excluding all or nearly all local access.

In our sites in China, villagers had more discretion over what local forests would look like. This was especially true under household-based management policies that granted local people extensive use rights to land, trees and forest products. At these sites, local people converted wasteland by planting non-timber species such as bamboo and fruit trees. They were also able to select their own timber species, generally Chinese fir or eucalyptus, when timber production was of interest to them. Local discretion was not, however, complete. Government officials exerted informal pressure on villagers to plant timber species in household plots that was sometimes difficult to resist. Collective forests were also subject to influence from the state. Where collective forest land was transferred to individuals and companies for fruit or rubber plantations, many villagers lost access to fuelwood, construction timber, medicinal plants and other important forest products.

Thus, from the perspective of the poorest forest users in all three countries, increases in forest cover often meant a reduction in forest quality. When regeneration was limited to a few species (often of little local value) that discouraged the growth of other species, or where regeneration was established on lands needed for grazing or agriculture, locals did not get the forests they wanted. Local initiatives to promote natural regeneration, particularly those in Orissa, did a better job of protecting species valued by the poor while still increasing forest cover, encouraging some timber production and protecting local soils and water sources.

## DIFFERENCES OVER WHO SHOULD BENEFIT FROM FORESTS AND HOW

Devolution policies provided direct benefits to at least some local forest users in virtually every case. These benefits took the form of improved access to subsistence products, improved access to forest product income, support for alternative livelihoods, and/or access to outside financial support. They also provided indirect benefits in the form of visibility, legitimacy and even legal recognition for local forest management initiatives, and the resources that such visibility attracts. Again, however, there are critical caveats to these findings. Firstly, many of the benefits promised did not materialize. The state reneged on its promises in many of our sites, leaving local forest users with little to show for early efforts to regenerate or protect forests. Secondly, the poorest forest users often did not receive a fair share, if any share, of the benefits available – a share consistent with their contributions in protecting forests or their dependence on forest resources for their livelihoods. Local elites and outsiders made sure that their needs were served first. Thirdly, local forest users often had little say over the terms under which they enjoyed these benefits. States offered limited options and locals often protested or worked around the terms, instead of accepting them. India's joint forest management (JFM) programmes were once again most disappointing in this regard, China's household-based management and tree sharing programmes the least, while the Philippines' community-based forest management (CBFM) programme fell somewhere in between.

On paper, devolution policies often grant local forest users significant rights to access subsistence products. As noted above, however, forests that fell under the influence of the JFM and CBFM policies were often created or transformed in ways that discouraged the growth of subsistence products. Legal access was undermined by reduced biodiversity and ecological destruction. In other cases, access to important subsistence products, such as fodder and fuelwood, was restricted in keeping with the management requirements of timber production or conservation. Fuelwood collection in parts of Orissa, for example, was restricted to just a few days during the year, ostensibly to protect the forest.

Policies were more miserly, even on paper, in granting local users rights to commercial species. Timber revenues, in particular, were largely reserved for the state and associated companies.[7] Local users had few commercial rights to timber in any of our three study countries. Only in China did several policies give communities rights to engage in timber trade. This trade was heavily regulated, however, and provided villagers with little influence over when and to whom they could sell.

Devolution policies often included support for activities designed to reduce dependence on the forest, such as generating employment or agricultural production outside of the forest (Sato, 2000). These have rarely made up for lost income and the value of subsistence goods and services from forests. Employment through reforestation schemes was too short term for villagers in Camarines Sur, the Philippines, to acquire capital, leaving them with little choice but to harvest resources from forests that they had agreed to protect. In Jinggu, China, villagers refused cash payments for lost access to forests, demanding rights to continue tapping resin instead. Forest users in Orissa also explicitly rejected the potential income from what they termed 'sharecropping arrangements' under JFM. States and forest users clearly differed as to whether and how livelihoods were interchangeable. Local forest users were interested in access to productive assets that supported local livelihoods, not just to income streams.

Even these limited local rights to earn income from forests were often not respected. JFM commitments regarding revenues from intermediate products have been routinely violated by the Forest Department in Orissa. In the

Philippines, the Department of Environment and Natural Resources (DENR) revoked CBFM permits for harvesting commercial products such as rattan, thereby limiting forest use to non-commercial uses. Not surprisingly, then, the limited income and employment associated with devolution policies was inadequate to cover the costs of local forest management in many communities. The disincentives of programmes were usually not immediately evident, but appeared over time with the implementation and maintenance of programmes.

Where projects disappointed local users, they often did not persist for long, as in some forms of collective forest management in China and early social forestry programmes in the Philippines. Still, JFM, CBFM and household management programmes have been accepted in large numbers of villages across India, the Philippines and China, respectively. Why is this so? As noted, in some cases, the disincentives to participating in a devolution programme were not immediately evident to local users, and opposition takes time to materialize. In other cases, the authority of government was difficult to resist. Forestry agencies maintained a quasi-military status in forest areas and had a strong influence. Opposition to a bad deal may appear more costly to local users than simply accepting further restrictions on forest access and use. This is especially true among the poorest and least organized of forest users.

Sometimes a sub-section of the local population, often the politically powerful, captured sufficient benefits to keep projects going, even in the face of opposition from the poorest forest users. In parts of Orissa, lower caste men and women were displaced in government-supported village protection committees by upper caste men who shared the forest department's interests in commercial forest species and forest conservation. In the Philippines, poor, less-educated villagers did not join CBFM organizations and profited little from them. In China, inequality increased as some households acquired much more forest land than others, based on greater levels of education and a willingness to take risks early when lands were available, but also on personal connections to government officials.

Outsiders managed to capture large shares of forest-related benefits, as well. Government officials were the primary beneficiaries of timber sales under government-sponsored shareholding systems in China. In self-initiated shareholding systems, farmers had more say over the terms of the contracts with outsiders, but were still taken advantage of – especially by forest companies with better information and influence in government. The harvesting of bamboo and other NTFPs was routinely contracted to private companies in Orissa, even after villagers invested years of labour in bringing the stands to the point of harvesting. Local officials and other elites received sufficient perks to ensure their complicity.

Devolution policies also provided notable indirect benefits to forest users, and these were sometimes important enough to encourage locals to continue regeneration and to protect forests in the absence of significant direct benefits. Across all three countries, local forest users gained greater visibility and legitimacy as a result of devolution policies.[8] Devolution policies encouraged more open discussion of management issues between foresters and local users, creating channels for the communication of local priorities to government decision-makers. Policies that granted formal rights to forest lands provided local forest users and their allies with a powerful political tool with which to confront those challenging their claims on the forest. The legitimacy of local forest users has increasingly been extended to their role in managing well-stocked forests and protected areas in all three countries.

This gain in visibility and legitimacy has enabled forest users to demand and gain increased attention from donors, government service providers and non-governmental organizations (NGOs). A flood of NGO, university and government-assistance programmes has occurred during the past 20 years to provide, among other things, legal literacy, community organizing, planting materials, nursery training, livestock, small enterprise development, agro-forestry techniques, environmental monitoring, and erosion-control training. Such attention has had its positive aspects in bringing new ideas, materials and projects to the communities. Women, in particular, may have received increased attention from these sorts of interventions because of the gender sensitivity of the organizations involved, especially in India and the Philippines.

There were problems, however, with even these indirect benefits. Many initiatives ignored what communities were already doing on their own, and have proven unsustainable over time or provided only partial solutions to a problem. Visibility has made it possible for forest users to be more easily 'counted' and thereby vulnerable to state regulation, as is the case of *van panchayats* in Uttarakhand. More of their resources have been inventoried and territories mapped, and the resulting information has been made available to government or other bodies.[9] In addition, policies that 'legitimate' local forest use and decision-making do not have the legal weight of formal legislation and can be easily undermined by legislative developments or court decisions. The benefits of visibility and legitimacy have therefore been mixed.

Devolution policies paved the way for greater benefits to accrue to local forest users. However, the direct income and livelihood benefits were smaller than hoped for, and often did not offset the costs of forest management. Local elites and outsiders, rather than the poorer forest users, often captured the largest share of local benefits. The poor benefited more from indirect benefits, especially visibility and recognition, though this was often a benefit that could be turned against them by government officials at a later date. In nearly all cases, the poorest local users rarely had much say over the terms of benefit-sharing and livelihood support.

## DIFFERENCES OVER WHO HAS THE RIGHT TO MAKE DECISIONS ABOUT FORESTS

Forest departments have maintained control over management decisions related to tree resources, work plans, budgets, market outlets and local organizations as strategic points of intervention. The application of such controls was most comprehensive in the Philippines and India. Government officials in each country controlled planning, supervised the budgets and decision-making processes of local organizations, and, in India, often controlled the marketing of forest products, including timber and the most commercially valuable NTFPs. In both countries, this represented not the maintenance of the status quo, but an actual loss of local *de facto* decision-making authority. Local organizations – whether self-initiated forest protection groups in Orissa, *van panchayats* in Uttarakhand or community-based organizations in the Philippines – that once made many management decisions on their own were subject to the oversight and discipline of the state.

In China, the trend was different. Far-reaching state control of planning processes was rolled back in favour of planning by households and communities. Though informal pressures on decision-making cannot be ignored, villagers were generally able to make many decisions about what to plant, where and when. Instead, taxes and regulations on harvesting and marketing timber were the most visible tools of state control. These indirect mechanisms were often no less burdensome to local users. This strategy of control has backfired somewhat on the Chinese government. Villagers have frequently opted to plant fruit trees or bamboo to avoid the onerous regulations associated with timber species, in spite of foresters' preference for timber plantations. Shareholding systems appear to be a response to this problem, as well as to the problem of land fragmentation. Villagers join with foresters to manage timber and let the foresters deal with the regulations and taxes.

In all three of our case study countries, states and local forest users seemed to differ over who had moral responsibility for, and claims to, forests (see also Peluso, 1992; Klooster, 2000). Foresters sometimes seemed to assume that they had moral authority over all forests. This was often phrased in terms of foresters' role in representing the 'public interest' (Lynch, 1998; Doornbos et al, 2000), though claims to technical expertise and the superiority of scientific management are also widely invoked (Scott, 1998; Guha, 2001; Malla, 2001).[10] Local users seemed to believe that such authority was theirs by virtue of the labour that they invested in forest management, their long-standing occupancy of forests, and/or their livelihood dependence on forests. This distinction was best illustrated by cases where local protection was initiated on degraded lands, only to have the state reclaim authority once forests had regenerated, amid loud protest from local users.

In much of Orissa, for example, foresters were happy enough to allow local forest users to plant and protect forests on what was nominally forest land, though such activities were supposed to be the responsibility of the state. Once forests had recovered, however, the state asserted its authority anew. It could not claim full control, as it did not have the resources to enforce such a claim. That is why it gave up trying to plant and protect in marginal areas in the first place. But it did demand a share of benefits from forests where it had invested little or nothing at all. The Forest Department's reassertion of control over regenerated forests extended to lands that were not included within the department's area of direct control – that is, over the *van panchayats* in what is now Uttarakhand and private forest lands in north-east India. Logging bans in the Philippines, China and India, abrogating a wide range of local rights and claims on forests, also provide evidence of the state's sense of moral authority.

In contrast, local users often claimed moral authority based on their efforts to protect forests. This was especially true in India, where poor men and women made great sacrifices to regenerate badly degraded forests. Local users worked without pay to patrol forests, plant, prune and do other maintenance work. They also devoted significant time and energy to organizing themselves to do this daily work and to make important decisions about management priorities. At the same time, many imposed limits on their own access to forest resources, sometimes for years at a time, to allow the forest to re-establish itself. Not surprisingly, these arguments bled into others, suggesting that locals were better managers than forest departments and should be allowed to carry on with their work.

Local users in India and the Philippines also spoke of their long history of forest management to justify their claims to the forest. In doing so, they highlighted the importance of customary practice in establishing moral authority, in opposition to the formal statutes and policies of the state. In some cases, however, local users could also cite policy and legal documents to support their claims to management authority, as with the *van panchayats* in India and the Indigenous Peoples Rights Act (IPRA) in the Philippines. Among indigenous peoples in China and the Philippines, important cultural beliefs associated with local management histories provided additional justification for local management control.

Finally, many local users claimed moral authority over forests by virtue of their dependence on forests for their livelihoods. In China, villagers were particularly effective in convincing state authorities to establish use rights for locals based on the role of forest resources in local livelihoods. Where authorities were unconvinced, as in much of India and the Philippines, a shared sense of local forest dependence seemed to legitimate ostensibly illegal activities in the eyes of neighbours and even some local authorities. Locals often recognized the needs of other stakeholders as legitimate, too, whether they were poor men and women from neighbouring villages, as in India, or state forest companies, as in China. Local needs, however, were the 'first charge' on forests in many villages.

In general, local users recognized the authority of the state in forest management. In fact, many argued that the state should do a better job of meeting its management responsibilities in forests that were allocated to them. But this moral authority was shared with local users. Where there was an extensive history of local forest management and continuing labour contributions, locals had the moral authority to continue managing as they had been. Where local users had both the capacity and the need to take up specific management tasks, they had the moral authority to do so.

## Strategies of state control

The outcomes above lead us to conclude that devolution policies represent *a shift in the manner in which central governments control forest management, rather than a genuine shift in authority to the poorest forest users*. Timber is still the focus of management practices in many forests (though conservation has become a second goal in places of particular ecological value); states and local elites still capture a large share of benefits from forest products; and states still exercise considerable direct control over forest management decisions.

What are the strategies by which the state has maintained its control over forests? Though government officials still engage in forest regeneration and protection in all three of our countries, they have increasingly transferred responsibilities for these activities to forest users. At the same time, they have retained authority over most other aspects of management, especially management planning and the disposition of forest products and services. In re-allocating responsibilities in this way, devolution has allowed the state to maintain or, ironically, even increase its influence over the day-to-day business of forest access and use. By supervising or manipulating planning processes, local organizations or tax and regulatory incentives, states exert control at far less cost, and can afford to reach forests where they had no effective presence before.

We now look in more detail at three important strategies that explain the outcomes above by showing how the state has maintained indirect control over forest management: taxes and regulations, contractual agreements and the supervision of nominally local institutions.

## Taxes and Regulations

Taxes and regulations are a necessary part of forest governance. They can be justified in terms of protecting public interests in ecological services and government revenues that fund planting, protection and other management activities. They can also work to support local forest users' efforts to manage forests (Lindsay, 1998) by restricting outsiders' access to local forests, using tax revenues to finance training programmes, or encouraging greater accountability of local government to the poorest forest users, among other things. Too often, however, taxes and regulations undermine local decision-making, with little evidence that they serve the public good. In these cases, we suggest that they represent a strategy on the part of the state to meet its own revenue goals and maintain control over forest management decision-making.

Taxes and fees distorted local forest management decision-making most dramatically in China, where significant tenure rights to trees have been devolved to individual households and households contracting with each other to grow, harvest and sell timber trees.[11] Taxes and fees on timber species discouraged farmers from planting Chinese fir, though that was their preferred species, rendering the farmers' rights to choose species much less meaningful.

Similarly, land development taxes imposed by local government officials in the Philippines have discouraged villagers from investing in social forestry. Ironically, this has encouraged illegal forest product harvesting in nearby natural forests. More to the point, it has also rendered the agreement virtually useless to local signatories. Such systems of taxation and fees are not likely to be good for government revenues in the long term (Daowei, 2001), but are certainly bad for farmers even in the short term, and reflect the state's continued focus on controlling timber production and harvesting and the revenues associated with them.

Regulations have also discouraged local forest users from planting timber species in China. Cutting and transportation permits, in particular, have proven onerous to villagers and have dissuaded them from planting timber species. The permits are granted based on quotas determined at the central level, first, then are filtered down through lower levels of government. Timber supply for urban areas must be assured, as must protection of watersheds that supply lowland urban and agricultural centres (Shen, 2001). Regulations are tightened or loosened depending upon which of these goals is at risk. These are, of course, legitimate goals. Once again, however, the regulations have been so strict as to discourage planting of timber species at all in many areas. As such, local tenure rights to timber species have little practical value.

In India, permitting is often arranged in coordination with local institutions that, as we discuss below (see 'Extending the arm of the state through local organizations'), are frequently controlled by the forest department anyway. It is worth noting here, however, that written permits are often required of local organizations not only to harvest and sell forest products, but also to get access to the income once it is earned. This was the case in many of the village forest joint management (VFJM) villages in Uttarakhand, and has been a focus of struggle in many JFM villages, as well. Villagers not only expressed frustration with not being able to obtain permits to harvest forest products to meet their needs, but also protested that permits *were* granted for illegitimate harvests by local elites working with forest department officials. Foresters retained substantial control over forest management – working around or through local organizations operating under JFM – by issuing or denying permits for a wide array of management activities.

## CONTRACTUAL AGREEMENTS

Much of the 'devolved' community-based forestry in the case study countries occurs through contractual agreements between public officials and forest-user organizations or individual households. JFM programmes in India, CBFM and social forestry in the Philippines, and shareholding and tree-cropping agreements in China all involve agreements that, in principle, commit signatories to certain allocations of rights and responsibilities in forest management. Such contracts should be appealing to officials and forest users alike. They could provide a clear division of roles among different forest users and officials, creating a climate of greater security in forest management. They could also provide legal mechanisms to challenge any violations of the rights of particular groups, and could hold each group accountable for meeting its responsibilities.[12]

On the positive side, contracts have opened space for forest users to make more decisions about forests in some circumstances. Generally, contracts that provided stronger land and tree tenure gave communities more flexibility and decision-making space than those confined to narrower benefits, such as shares to a product harvested. Villagers have also borne more risk economically where contracts awarded only a share of benefits, rather than rights to a productive asset. In this sense, devolution policies such as JFM in India have been the most restrictive. CBFM and social forestry programmes in the Philippines are more progressive in granting use rights over land and trees, generally in the form of 25- and 30-year renewable leases. In the Philippines, however, the lease terms, especially requirements for forest protection and the planting of certain species, can be quite restrictive (see Li, 1999). Perhaps the most space has been created in China where, under tree-sharecropping systems, farmers could contract with each other over forest uses, rather than with the state. Though the contracts were subject to other state regulations, as well as to informal pressures from the state, the forest users themselves worked out how to share responsibilities and rights in these contracts.

In most circumstances, however, the contracts have not been favourable to forest users. Devolution policies in the case countries have not addressed the need for greater accountability of all forest interest groups to one another.[13] Instead, the policies have, to varying degrees, reflected forest departments' needs to reduce initial investments, create greater management flexibility, and still preserve central control over many of the key decisions to be made about forests. In all of the case study countries, governments used contracts to encourage people to undertake the labour-intensive tasks of forest management – planting, cleaning, protecting and even harvesting – while restricting local rights to make decisions about how to manage forests and what to do with forest products.

Forestry officials have used contracts to maintain control in several ways. Firstly, many of the contracts reflect a forester's vision of what constitutes a good forest, as discussed above. For example, shareholding contracts in China and most JFM contracts emphasized timber production, while saying relatively little about fuelwood collection, access to fodder, collection of medicinal plants or wild foods or other non-timber uses of the forest. Forest users in the Philippines have resorted, in some places, to 'illegal' use of forests at the risk of fines and other punishments from foresters because legal recourse was unknown or unavailable to them, despite the existence of the contract.

Secondly, contracts have defined the responsibilities of users and the state in ways that reflect state constructs and interests. In Orissa, foresters promoted social forestry programmes in which they would select tree species, organize nursery development and transplanting, and regulate harvesting and marketing, instead of enforcing locally defined forest access and use regulations, as many villagers had hoped. Even where local institutions were formally recognized in contracts, as in CBFM programmes in the Philippines and JFM in India, planting, harvesting and marketing decisions were often predetermined by forest departments, often in standardized terms inappropriate to local ecological, social or historical contexts. Moreover, contracts often built in technical advisory and managerial roles for foresters that created opportunities for officials to charge (formal and informal) fees for services. Chinese farmers who signed shareholding contracts, for example, sometimes were pressured to accept species and management regimes for which foresters had special technical expertise, and to accept a regulatory system that required farmers to obtain cutting and transport permits, pay taxes and otherwise provide foresters with substantial opportunities for income.

Thirdly, contracts were approved and dissolved by forest departments with little accountability to forest users. As a party to the contract, it is not surprising that the state would need to approve the terms before signing any contractual agreement. As has been reported for India, China and the Philippines, however, the terms of contracts were usually not widely discussed among forest users, and awareness of their terms was limited. In these circumstances, contracts acted as a disciplinary tool for forest users, legitimating state interventions that were never fully understood or accepted by large numbers of forest users. Evidence from Uttarakhand suggests that several communities felt deceived by the terms they, in theory, were contracted to respect under the VFJM programme. The contracts limit existing forest access and decision-making rights in a way that was not clearly explained, even to the small number of villagers who were actually consulted. Communities in Gengma, China, also felt pressured to accept contracts proposed by township and county officials for shareholding arrangements. The arrangements asked forest users to bear substantial risks for uncertain benefits, while providing them with little say in day-to-day forest management.

At the same time, the state broke off agreements in ways that dramatically disrupted local livelihoods. The worst examples were the logging bans in north-east India and western China that effectively abrogated local use rights over large and diverse landscapes. The bans were justified in the name of poorly defined, sweeping allegations of forest abuse by forest users, and punished households and communities who had been protecting forests on terms, in principle, approved by the state. Under the CBFM programme in the Philippines, the state also repealed use rights to forests to support local livelihoods when local people committed minor violations of the forest protection guidelines. Those who had protected in good faith again had little recourse against this collective and arbitrary punishment.

Finally, contracts were simply violated by public officials. As noted, foresters in Orissa routinely ignored the claims of forest users to all intermediary produce, as stipulated in their JFM agreements. Officials contracted state companies (often private companies with good connections to the state) to manage NTFP harvesting and marketing in violation of the agreements. In Gengma, contracts promising payments to forest users for the use of collective land were not made, whether or not the plantation projects established on collective land failed. In each case, forest users found little relief in complaints to government officials, even in court. Again, in these conditions, contracts appear to be instruments of state control, rather than agreements of mutual accountability.

Contractual agreements are easily the centre-pieces of devolution programmes in India, China and the Philippines. Unfortunately, these instruments have continued to reinforce many of the constructs and interests of the forest departments. There is, nevertheless, substantial promise in them, as can be seen from the case of some of the tree-sharecropping arrangements in China. Further progress is needed to ensure that the poorest farmers have access to advice on negotiating contracts and on legal protections when contracts are violated or conditions change, as well as the capacity to enforce contracts and legal remedies when necessary, even against the state. Until these conditions are met, contracts are better understood as a mechanism for controlling local forest users rather than empowering them.

## Extending the Arm of the State through Local Organizations

Devolution generally requires some legally recognized target institution to take on management authority. Questions have emerged, however, over when and how to create new organizations, instead of relying on those that already exist in a community (Gilmore and Fisher, 1997; Mandondo, 2000); whether the state should be the prime mover in such a process (Evans, 1996; Harriss and de Renzio, 1997); how to organize the accountability of local organizations (Ribot, 1999); and how local organizations interact with other interests to shape management decisions (Anderson et al, 1999; Wollenberg et al, 2001); among others. In our case study countries, the answers to these questions have generally been shaped by the constructs and interests of forestry officials at the expense of local forest users.[14]

Community-based organizations in our case study countries have had the potential to help forest users wield influence over forest departments, take advantage of economies of scale and foster greater democracy in forest management by developing inclusive and transparent decision-making processes (Arnold, 1989; Colchester, 1994; Ostrom, 1999; but see Agrawal and Gibson, 1999). They have made progress towards reaching this potential in some areas, especially among the self-initiated forest protection groups in Orissa, the most successful CBFM sites in the Philippines, and those village committees in China that enjoy significant popular support. Where they have not met their potential, the fault can often be found among local forest users themselves. Local elites have manipulated local organizations to at least some degree in the majority of sites in all of our case study countries (see also Sarin, 1998; McCarthy, 2000b; Gauld, 2000; Malla, 2001). A lack of managerial experience – in marketing products, adjudicating disputes, planning a harvest – has sometimes hindered local organizational success, as well (see Bebbington, 1997; Richards, 1997; Brown and Rosendo, 2000). Nevertheless, one of the most important reasons for local organizational failure is that officials have played a dominant role in determining how they have functioned, and have steered them in directions favouring forest department constructs and interests to the detriment of local livelihoods and decision-making authority.

One of the problems is that the state has often created new organizations that compete with or replace existing decision-making institutions at the local level. In India and the Philippines, forestry officials have been the driving agents behind developing new organizations as conduits for forest devolution programmes – namely, through protection committees in India and farmers' associations in the Philippines. These are often layered over locally initiated institutions that are more deeply embedded in local social relations, cultures and environments. Although new organizations can sometimes address local inequalities, they often reinforce them. When new institutions are formalized and conduct business in a bureaucratic fashion and under the gaze of public officials, they can effectively exclude large sections of forest users (see Gauld, 2000; Sundar, 2000b; Malla, 2001), as has been the case with JFM committees, CBFM organizations and even shareholding groups in China.

By introducing new organizations, foresters can also re-orient organizational objectives. Forest protection that preserves local livelihoods thus becomes protection for the sake of protection. In the Philippines, the Palawan Regional Council seized decision-making initiative from a wide array of informal decision-making institutions, and livelihoods are now at risk from stringent conservation measures. Efforts to form local organizations may also be token gestures to fulfill bureaucratic requirements. The Bicol National Park case in the Philippines demonstrated how a flurry of social organizing occurred to produce seven new people's organizations, none of which addressed the fundamental problem of tenurial rights in the buffer zone.

Where the state felt new institutions were needed, as seems to have been the case in much of rural China, there is a question as to whether the state should take an active role in forming them or merely encourage nascent organizational development by communities themselves (see Mandondo, 2000; Jenkins and Goetz, 1999). In several of our case sites in China, locally initiated tree-shareholding groups have done well in comparison to state-initiated shareholding systems. Farmers exercise more control over decision-making in the former, including the terms of labour and benefit-sharing under contracts. Though the evidence is not clear, there appear to be greater concerns about coercion and corruption in government-initiated shareholding systems, as well (Song et al, 1997).

In our case study sites, states frequently made use of existing local organizations. Problems of co-optation arose, however, which made these organizations no less state constructs than those directly initiated by the state. Joint forest management in India and community-based forest management in the Philippines required local organizations to register with the state and have management plans approved by forestry department officials.[15] Indian foresters sat on all protection committees organized through JFM, often in positions of decisive decision-making authority, but had no formal accountability to forest users. Where local organizations were absorbed within JFM, villagers complained about the re-orientation of priorities towards timber and/or conservation, and about the narrowed scope of their authority over their forests. In India, the forest department even asserted control over money saved by villagers *prior to* the implementation of any devolution policy. Many local forest users also pointed out that procedural rules under JFM diminished the participation of disadvantaged groups in decision-making, especially poorer women, through literacy requirements and other means.

In China, the lines of accountability were somewhat more opaque and, relative to earlier days, freer of state interference. Nevertheless, village committee members were elected only after party officials approved their nominations. Though members did not report directly to party officials on normal decision-making, there was always the threat of removal for decisions against the party's interests. In all of our case study countries, an informal accountability to forest users sometimes occurred, particularly where individual foresters were sympathetic to local claims on forests. Senior officials, however, had the option of circumventing this kind of relationship through formal bureaucratic mechanisms when the need arose. Thus, in the Philippines and India, popular local foresters who were committed to working with, rather than over, forest users were transferred out of their areas.

Finally, local organizations are often marginalized in agreements with other stakeholders (see Gupta, 1995; Bebbington, 1998a; Tiffen and Zadek, 1998; Brown and Rosendo, 2000). In both India and the Philippines, for example, forestry and other departments signed agreements with private companies to exploit lands where community-based organizations were managing forests, and did so without consulting local organizations. Plans to protect forests as conservation areas were also developed without input from local organizations in India, China and the Philippines. The failure to consult local organizations on matters of such critical importance to them betrays an official disregard for local rights. Local institutions appear, instead, to be a means of perpetuating state control of forests by tapping the labour and social capital of local people. From this perspective, state approaches to local organizational development recall strategies of 'indirect rule' of the colonial era, where devolution maintains central control, but at lower administrative and political costs (Mandondo, 2000).

## CONCLUSION

The constructs and interests of state officials diverge in important ways from those of many of the poorest local forest users. Specifically, state officials and these local users often differ in their views of what counts as quality forest, who has rights to what benefits from the forest, and who has the authority to make decisions about forest management. Because the state maintains control over policy design and, to a lesser degree, implementation, devolution policies have, by and large, reflected the constructs and interests of the state. As a result, devolution policies have often worked against the constructs and interests of local forest users, particularly the poorest among them.

The state has employed a variety of strategies to overcome local opposition to its policies. Though some of these involve direct management of forests by forest departments, states increasingly make use of strategies for maintaining indirect control over forests. These include taxation and regulation, management contracts and the manipulation of local organizations. In some places, these strategies have been effective in maintaining state control, often at the expense of the poorest forest users. However, there are examples from our case studies where state and local interests converge. There are also examples where local interests have prevailed in spite of state strategies to maintain control. In Chapter 6, we look at these examples of convergence in more detail and make recommendations for how a meaningful devolution – one that respects the constructs and interests of the poorest local forest users – can be achieved.

# 6

# Conclusion

*David Edmunds, Eva Wollenberg, Antonio P Contreras,
Liu Dachang, Govind Kelkar, Dev Nathan, Madhu Sarin and
Neera M Singh*

We have argued that devolution policies in our case sites have reflected the conceptual frameworks and interests of foresters and, as a result, have disappointed local forest users with different expectations of devolution. The state in each of the three case study countries has maintained control over forests through taxes and regulation, contractual agreements and influence over local organizations. The divergence between the state's interests and those of local users reflects a clear trend of state control across the three countries. The space for local forest management is larger according to governments. According to many forest users, it is not yet large enough to make a difference to their livelihoods.

Hence, in China, local users have been able to make important management decisions, capture significant benefits and create forests according to their own preferences, especially in household-based management and some shareholding systems. However, the state continues to exert control over timber production through taxes and regulations, and officials exert informal pressure on lease and shareholding agreements for all forest land uses. In the Philippines, local users have benefited from agroforestry, reforestation and forest protection projects. Yet, the Department of Environment and Natural Resources (DENR) defines what the forests will look like, what benefits will be available and on what terms locals will gain access to them, with little effective input from local users. The Indigenous Peoples Rights Act (IPRA) holds promise of more meaningful devolution, where local users will make most of the important decisions about forest management. However, the implementation of the act is still in its earliest stages, and it is difficult to know what limits the state may try to place on local decision-making. The situation is perhaps worst in India, especially where local people were already organized to manage forests. In these areas, local institutions for managing forests, some formally recognized by the state, have been undermined by Joint Forest Management (JFM) and related state initiatives. Decisions once made at the local level to protect and regenerate forests to meet local needs are now made under the influence of foresters and their allies. For many of the poorest forest users, this has meant a loss of livelihoods and increased threats of fines, imprisonment and violence.

Chapter 5 presented a bleak assessment of the current state of devolution policies. There are situations, however, where policies have helped to create limited space for local users to influence natural resource management. Many of these represent situations where the interests of local users converge with those of public officials. Examples are too rare, but are still significant (see Wollenberg et al, 2001, for a general discussion). There are also a number of areas of near convergence that have not yet been developed, but could be as the social and political context for natural resource management continues to change. In this chapter, we highlight where we think convergence has already occurred or might occur in the near future. We also discuss efforts by forest users and their allies to create new space for local decision-making, either by working around and against policies, or by reforming them. We describe some of the specific strategies employed and indicate some of the strengths and weaknesses of each. We end with general recommendations for expanding the space for local decision-making in natural resource management.

## WHERE CONSTRUCTS AND INTERESTS CONVERGE

Although the constructs and interests of forest agencies and local people generally differ with regard to forest use and management, this is not always the case. In general, we found that convergence occurred in three ways.

Firstly, forest users' assessments of their role in decision-making depended, in part, on what their role had been previously. That is, users were sensitive to relative improvements in their condition. Where forest agencies had previously managed forests in highly exclusionary and centralized ways, devolution policies sometimes made it possible for villages to increase their overall control over decision-making and to use forests to meet more of their own needs.[1] This was truer in China than anywhere else, where local decision-making had been severely limited prior to the start of the reform period in 1978. This partly explains the generally positive assessment of devolution policy impacts there. Where forest agencies had low presence or failed to control use, which was more common in parts of the Philippines and India, local people developed their own *de facto* policies for managing forests, and 'devolution' policies intruded on existing practices. Context varied within states, as well as between them. In India and the Philippines, social and community forestry programmes during the 1970s and 1980s were viewed by many as 'win–win' outcomes that met the state's aims for increased forest cover and local people's livelihood needs (Poffenberger 1990). The projects were more likely to be evaluated as successful, however, in degraded areas and areas where no local initiatives were already in place (Arnold, 1989).

We found, however, that even where there may have been initial feelings of progress, local forest users were rarely satisfied with the level of local decision-making for very long. This was especially true as forests became more valuable and as the allocation of labour and benefits among different groups within communities looked increasingly unfair. Thus, in China, where tenure reforms such as the household responsibility system were quite sweeping and enjoyed much popular support, people soon complained about, and sought to circumvent, the burdensome tax and regulation system that had been developed for timber harvesting. In India, poor women participating in JFM agreements soon recognized that they were contributing substantial labour to forest protection to ensure supplies of timber or commercial non-timber forest products (NTFPs), not the forest products that they needed (Sarin, 1998). In the Philippines, many communities shifted their focus from community-based forest management (CBFM) to attaining more far-reaching management rights under the IPRA. In the worst cases, local elites or government officials usurped the benefits of forest management once resources became valuable. Women from poorer households in Orissa, for example, have left JFM committees since wealthy male farmers seized leadership positions after trees began to regenerate. Convergence has been achieved, but has often proven to be fragile.

Secondly, convergence was more likely to occur where local people and government officials divided roles and responsibilities in ways that enabled local people to make their own day-to-day livelihood choices with a maximum of discretion, while the state provided support for these choices and controlled the quality of public good outputs. In China, farmers were granted extensive rights to plant, maintain, harvest and sell bamboo as they saw fit. Incomes rose substantially, however, only when officials from local government provided access to first-stage bamboo processing equipment, organized technical advice on bamboo management, and secured funds to build a factory nearby for making floor tiles and other products from bamboo. The local government has benefited from increased revenues from taxes on bamboo, and households enjoy much more disposable income, as well as greater discretion over the bamboo management. There was a high degree of dependence on the state-owned factory, in this case, which raises concerns about how much choice farmers really enjoy in marketing their bamboo. Nevertheless, farmers who were interviewed appeared quite happy with recent developments in their village.

In other areas of China, tree sharecropping has also put the state in the position of supporting the decisions made by local forest users, rather than constraining them. In the most promising cases, associated primarily with household-based management, locals are allowed to grow trees and use them as they wish, or to lease trees out to others to manage if they do not have the time or interest to do the work themselves. When leases are arranged, resource inputs, management responsibilities and prices are negotiated in advance, allowing villagers to exercise considerable authority over their own role in management. Government officials provide technical advice and help to enforce contract terms, while banks and companies may provide capital for management activities. Of course, to the extent that any of the terms of the lease are forced on farmers, or farmers are compelled to lease land against their wishes, the state continues to exercise authority much as it did prior to the development of tree sharecropping. This does not seem to be the case in most of the sites studied in China, however, and makes tree sharecropping a promising area for future development.

In some cases, villagers with more extensive tenure rights over forests have been able to decide when and for what purposes foresters will be used as technical resources. This was the case in some of the *van panchayats* in Uttarakhand, where foresters were expelled from local decision-making bodies early in the century, but could be called on by *van panchayat* committees to offer assistance on technical matters as needed. In theory, foresters were legally bound to provide such assistance, though the reality was complicated by all of the informal relationships that determine the effective accountability of foresters. This arrangement is being reversed with the absorption of the *van panchayat* committees by JFM, to the disappointment of many local forest users. Filipino foresters have also placed themselves at the service of local forests users in some CBFM sites, particularly in sites where local foresters have strong personal ties to the community and/or are sympathetic to the interests of poorer forest users for ideological reasons. Again, how much control communities exercise in these sites is unclear; but the principle is one that local forest users can warmly embrace.[2]

Where states have allowed more decision-making space to local institutions, they have sometimes also taken up a greater role in dealing with oppressive local social relations as they relate to forest management decision-making and benefit-sharing. While this certainly does not please everyone within forest-using communities, there is some convergence of interests with politically disadvantaged groups. Forest departments in India, under pressure from various popular and professional activists, have begun insisting upon the inclusion of women within village protection committees and other local institutions, and upon respect for women's uses of forests in village management plans (Khare et al, 2000). Again, the shift should not be overstated, as questions of whether women *effectively* participate in local decision-making remain (see Sarin, 1998, for a critical assessment). In the Philippines, state recognition that indigenous peoples have been politically marginalized on both local and national scales has contributed to the development of the IPRA. The act brings together a remarkable alliance of government officials, indigenous activists and non-governmental organizations (NGOs) to fight for devolution of governance (including rights to manage resources) to indigenous peoples and their organizations. Here, too, there are questions about state motivations and capacities to implement the act. Nevertheless, that states should take some role in addressing local forms of oppression, even as they retreat from other aspects of management, holds substantial appeal to disadvantaged local users (Lele, 1998).

State officials and local users can sometimes cooperate over the issue of protecting forests for various ecological services. Certainly, state involvement in forestry is increasingly justified in terms of environmental protections rather than production targets (Guha, 2001), as the cases from India, China and the

Philippines illustrate. We might question the qualification of officials for this role, given the history of overexploitation of state-managed forests in all three countries, and the continued ambiguous environmental effects of recent logging restrictions in China and India (Hyde, 2001). The frequent tension between state-organized conservation and local livelihoods would also seem to make this an unlikely point of convergence (Scott, 1998), as reported in the work on the Philippines and India. Where states assume the role of 'enforcer of last resort', however, helping local institutions to enforce locally designed and monitored restrictions on forest conversion, they can find broad local acceptance. This seems to have been the case historically in Uttarakhand, where *van panchayats* could call on the forest department for such help.[3] If states can take up the role of 'enforcer of last resort', their role in forest protection can complement, rather than contradict, local interests.

Another important point of potential convergence over roles and responsibilities is in the response to private companies, wealthy individual investors and other capitalists in forested areas. Local forest users may seek agreements and cooperation with such interests (Salafsky et al, 1999; Tiffen and Zadek, 1998) in the hopes of securing jobs and income. As we have seen in several of our case sites in China, local users need the capital and technical expertise that private interests possess, and are willing to sign agreements with these interests to provide a return for their investments. Many of the existing agreements have been facilitated, in part, by government officials.[4] States can thus help to bring private capital and local forest users together for forest development projects. As the cases from Gengma illustrate, however, local officials do not always facilitate agreements that are beneficial to local users.

In fact, it is probably more common that local forest users need protection from mining and timber companies, hydroelectric companies or even organized traders in NTFP seeking to exploit local resources at the expense of local livelihoods (Hecht and Cockburn, 1989). Unfortunately, there are few examples where governments have shown much interest in defending local rights against such 'development' projects. In Kashipur, Orissa, women have struggled to obtain rights to broom grass. Their rights mean little, however, as bauxite mining in the area is converting forests into virtual livelihood deserts. In Madhya Pradesh, NTFP collectors face petty traders who capture most of the profit from NTFP sales. In this case, the state government did try to intervene to help collectors, in part to ensure the provision of supplies and revenues, but also to protect local collectors. Local politics, however, limited the ability of the poorest collectors to take advantage of government services. NTFP collectors depended upon petty traders for a wide range of credit and supply services, and were afraid to alienate them by working with the government-sponsored co-operative. Local politics of this sort make state intervention complex, even where the government shares an interest in reforms with poorer forest users against the interests of private corporations.

# EXPANDING THE SPACE FOR LOCAL DECISION-MAKING

The previous section outlined areas of convergence or potential convergence between the constructs and interests of the state and those of local forest users, especially the poorest among them. This, however, begs the question: how do such areas of convergence develop? Or, more instrumentally: how can forest users and their allies – researchers, activists and officials – help them to develop?

We can look for existing areas of convergence as fruitful areas for investment, organizing efforts and policy reforms in forestry. Unfortunately, these areas may be few and unstable (Anderson et al, 1999; Sundar, 2000a,b; Baviskar, 2001). Given the power dynamics in our case study countries, this will leave large numbers of the poorest forest users with little space for making decisions about forests, their management and the use of forest products. We therefore look at some of the conditions and strategies that enable the poorest forest users to create areas of 'convergence' by changing the power dynamics with respect to the forests that they use. These strategies fall into two inter-related categories. Firstly, local users can cooperate – among themselves at the local level and through alliances with outsiders that share or will support their constructs and interests in forests. Such cooperation depends upon the inheritance and creation of social capital at various scales (Evans, 1996; Harriss and de Renzio, 1997). Secondly, local users can pressure other groups to respect their constructs and interests in the forest through lobbying, protesting and other political activities.[5] Forest users in our case study countries used strategies of cooperation and pressure, often simultaneously, to protect and expand the kinds of decisions that they wanted to make regarding forests. In fact, cooperation was often a requirement for effective protest or lobbying, and protest sometimes created new alliances for local forest users.

As many of our cases in India and the Philippines demonstrated, forest users organized themselves into user groups in order to take back some measure of decision-making authority lost to the state during the colonial period, particularly when the state appeared incapable of exercising effective control over the forests itself (Arnold, 1989). The groups were often formally organized, with clear rules of access, mechanisms of enforcement and decision-making procedures (Ostrom, 1999), even if they were not recognized by the state. Indigenous peoples, with norms of trust developed through myriad shared cultural practices, often had very effective local institutions for managing resources, as in Libo and Lijiang, China, and Nueva Vizcaya, Sarangani and Mindoro Oriental, the Philippines. Forest users also benefited when they had experience with other forms of local organization, such as labour exchange networks or rotational credit societies, that could be applied to organizing for forest management. In contrast, in some areas of China, negative experiences with collective forest management prior to the reform era discouraged local users from initiating formal organizations to coordinate forest use. Similarly, where states co-opted local initiatives – in Orissa and Madhya Pradesh, India, and Bukidnon and Mindoro Oriental, the Philippines – social capital has been eroded and local organizing is now more difficult than it was before. This is especially sad where local organizations had performed a good job of regenerating forests that were badly degraded under state management (Sarin, 1998).

In such cases, local social capital and organizing may need to be nurtured by outsiders. At a minimum, the state can play a role by improving the legal environment for local organizing (Lindsay, 1998). Forestry officials have also played mediating roles in some cases, helping to bring villagers together in formal organizations, providing guidance on decision-making procedures, and offering financial and technical support. Unfortunately, most foresters are not trained as facilitators and have few skills for the role (Mayers and Bass, 1999). More importantly, as we have suggested, the temptation for officials to advance state interests is great, and such facilitated organizations often work against the interests of the poorest forest users. This applies to organizations initiated by 'service-oriented' NGOs, as well. These NGOs often depend upon the state for their legal standing and financial solvency, and appear loathe to challenge state interests in forestry (see Bebbington, 1998, and Sen, 1999, for general discussions). Still, in all three of our case study countries, foresters, other public officials and NGOs have been instrumental in organizing forest users in places where social capital appeared to be scarce – often as a result of past abuses by forest departments and others – and the need for local organization great (see also Saxena, 1997; Guha, 2000).

Another caveat is needed with respect to local social capital and local organizing. As has been noted by others (Hariss and de Renzio, 1997), social capital among a set of local people often implies strict exclusion of those outside the group. This has led to equity problems in some of our case study sites. As noted, strong cooperation among the Naxi in Lijiang has allowed them to orient local forest management towards the needs of the tourist industry, at the expense of Yi people more directly dependent upon forests for subsistence and income. The earliest villages to organize in Orissa and Madhya Pradesh have sometimes abrogated long-standing use rights of neighbours (though generally under pressure from forest department officials working within a JFM framework). Nevertheless, the need to organize is clear. Even where villagers have been intensely sceptical of organizing, as in our sites in China where household management is prevalent, cooperation among households is emerging in the form of lease agreements, pooled-investments coordinated marketing and similar activities.

Improved coordination among local forest user groups and efforts to create mass movements have also expanded local decision-making authority and protected local livelihoods (Gilmore and Fisher, 1997; Froehling, 1997; Parajuli, 1998; Mayers and Bass, 1999; Bray, 2000). Federations and NGO allies in India and the Philippines have been particularly effective lobbyists for policy reform (see also Down to Earth, 2000). In Orissa, a mass protest over NTFP policies led to recent changes in the kinds of NTFPs regulated by the state, opening up space for local management for the first time since nationalization of NTFPs.

In Libo, China, farmers organized as a tree-sharecropping group brought court cases and organized media events to call attention to threats to their forest-based livelihoods. Ironically, the very standardized policies that imposed the state's vision of forest management on a wide array of local communities with diverse environmental and social conditions helped to create these super-local organizations. In other words, a common struggle against the state encouraged forest users from diverse villages to see themselves in a common light and to work together (Evans, 1996). Again, such super-local organizations can be co-opted by the state or elites. Chapter 3 on India provides an example. Donors have met public consultation goals conveniently and inexpensively by convening federation meetings. Villagers, however, complain that their representatives no longer voice local concerns over donor programmes (see Malla, 2001, for an example from Nepal). In fact, co-optation may be a bigger problem for federations than for local organizations, given that representation is more complex at larger scales and representatives move in different social and professional circles (Melucci, 1996). Nevertheless, super-local organizing can be a powerful mechanism for diminishing power imbalances between the state and the poorest forest users; in doing so, it expands the space for local decision-making.

Alliances are often formed between local forest user groups and others with less day-to-day contact with, or dependence upon, the forest. Forest user groups have allied with international conservation organizations (Brown and Rosendo, 2000; McCarthy, 2000), donors (Malla, 2001), local NGOs (Sen, 1999) and sympathetic government officials (Gupta, 1995; Wily, 1997; Moore and Joshi, 1999; Alsop et al, 2000), among others. An alliance of indigenous peoples' organizations, local NGOs and key professionals within the Philippine government has pushed through the IPRA, despite significant opposition from many Filipino foresters. Sympathetic government officials joined NTFP collectors and local activist organizations to push for the development of co-operatives in Madhya Pradesh, India. Local NGOs have also worked with forest users in legal literacy campaigns in the Philippines and India, raising awareness of forest user rights under various devolution policies. Foresters and NGO staff have provided useful technical assistance in several of our case sites, as well, related to species selection, harvesting schedules and other aspects of forest management and product marketing. Local governments are also potential allies for disadvantaged forest users. The adaptation of national policy to local circumstances by local government in China helped to provide greater benefits from protected areas in Yunnan, especially resin tapping, and allowed shareholding systems to develop throughout the south-west of the country. Local government institutions have also been empowered to manage forest resources for the benefits of local users under the *van panchayat* scheme in Uttarakhand and under the Panchayats (Extension to the Scheduled Areas) Act, 1996 (PESA) in Madhya Pradesh and Orissa, although to exercise that power is often difficult.

Such alliances can reproduce power imbalances, rather than challenge them. Shareholding agreements in China, for example, have often left little room for villagers to make decisions about the trees that they are supposed to own – such decisions are made by their government partners. Many NGOs, nominally allied with local forest users, have also reinforced state control over forest decision-making, perhaps unwittingly. NGOs and donors worked together to design JFM committees and CBFM organizations that were more accountable to foresters than to forest users. NGOs throughout the Philippines also have been instrumental in drawing up forest management agreements that restrict local access to the forest and threaten the livelihoods that depend upon that access (see also Ramos, 1994; Brown and Rosendo, 2000). Pressures to finance local government activities in the Philippines have also placed local government in a competitive position with local users, with both interests looking to increase their shares of income from forest products. These examples indicate clearly that forest users must be careful in choosing their allies. They have little choice in most circumstances, however, but to seek allies outside of their local communities if they are to exert influence over how forests and forest products are managed.

## LOOKING AHEAD

Although we are discouraged by what we have found in our study, there is room for optimism. Policy can work to the benefit of local forest users where local and state interests already converge, as in some income-generating and conservation projects. Local users have also developed alternatives to state-driven community forest management that better meet local needs. Yet, far-reaching change in forest policy is still necessary if forests, forest-related benefits and forest management decision-making are to serve the interests of the majority of local forest users, particularly the poorest among them. This change is likely to come about as the result of the political work of forest users themselves, organizing at the local level, establishing networks with other local organizations, and building alliances with sympathetic government officials, private capital, NGOs, donors and researchers.

This does not mean that others involved in forest policy can or should become inactive. Local users are not a homogenous group (Enters and Anderson, 1999), and those who work with them must be able to identify those users most in need of support. After years of relative isolation from formal, legal decision-making processes, many forest users may not be able or willing to communicate their interests to others effectively. Government officials, NGOs, donors and others will have to develop communication channels that forest users feel comfortable with and will use (Engel et al, 2001; Steins and Edwards, 1999). These same 'outsiders' may also have a role to play in building local institutional and technical capacities, though what this role should be is not always clear.[6]

## Lessons learned

We still believe in the potential of devolution policies to bring about better forest management, with more democratic decision-making and improved support for local livelihoods. For this to happen, many of the current policies will have to be repealed, replaced or drastically reformed. Local forest users are already working to bring about the necessary changes. In all three countries, they are creating alternatives to the management systems currently favoured by the state – new shareholding arrangements in China and self-initiated user groups in the Philippines and India. Local users are also organizing to demand policy reform, working through federations of forest users groups, NGO networks and even village committees. None of these actions implies that the state will be thrown out of forest management, or that there aren't legitimate state interests in forests that must be accommodated. They simply reflect the fact that, despite rhetoric to the contrary, many current policies leave little space for local users to express and act upon their interests, and that this has created often unbearable hardships for them. We support these efforts and hope to see more develop.

The audience for this book, however, is not likely to be forest users themselves. We have, instead, written the book for government officials, researchers, donor organizations, technical advisers, NGOs, activists and others whose work, like our own, affects or is affected by devolution policies and local efforts to change those policies. In closing, we offer some of the lessons we learned during the course of our research that will affect how we work on forestry issues. The lessons fall into six broad categories; but each is related to the others, and we hope to apply them in a holistic way to the work we do.

## Start with what forest users know and do

Many of the most successful examples of local forest management, in terms of regenerating and protecting forests and supporting local livelihoods, are found where forest users themselves initiated a management system. Unfortunately, central and state governments have often either ignored local initiatives or tried to bring them under a standardized framework for bureaucratic convenience and to re-orient management to state priorities for timber production or conservation. This has been particularly true of JFM in India and social forestry in the Philippines. We should, instead, start with management that is present on the ground and build from there to meet public interests. This is a daunting challenge for those working at the policy level, but not an insurmountable one. Mayers and Bass (1999) suggest that simple policies that allow for maximum flexibility in implementation can better accommodate a wide variety of local conditions, including:

- management priorities, usually linked to specific livelihoods in which forest users are heavily invested;

- institutional arrangements, taking into account existing imbalances of local power and interest, and the present state of alliances and antagonisms;
- divisions of roles and responsibilities that take advantage of differential skills and knowledge.

Allowances for the local formulation of bylaws, allocation of resources and personnel, and development of management activities are important aspects of flexible policy. Flexibility also implies that our work should evolve as social and environmental conditions change. This may mean foregoing long-term, tightly worded contracts between the state and local users. Instead, we can support collaborative monitoring of policy effects by governments and forest users.

Perhaps surprisingly, China has done the best job at this. The central government generally permits townships and administrative villages either to develop their own 'experimental' forms of forest management, such as self-initiated shareholding systems, or to decide which of the other management forms (collective, household, shareholding) are best suited to local conditions. Provided decisions fit within a broad framework of government forestry objectives, there is relatively little interference. Local initiatives are monitored for their effects and, if successful, are allowed to spread to other areas. This approach is one that deserves consideration in other places and strengthening within China itself.

Of course, existing management forms will have problems that may require outside intervention. Concerns for equity, accountability and forest quality can be particularly important in this regard. We can address these issues by evaluating the outputs of forest management, such as forest quality and equity indicators, rather than by controlling day-to-day decision-making. We can also support downward accountability and other process standards among local organizations. We discuss this further below.

## Create opportunities for pluralistic decision-making and provide disadvantaged forest user groups with the means to influence policy

The days of hegemonic control by forest departments are gone, and there is increasing recognition of the need for pluralism in the ways that forests are managed. As we learned, however, forestry departments have often been able to contain local influence over forest policy and practice by building alliances with local elite, controlling the flow of information and financial resources, and even resorting to intimidation and violence. Mechanisms are needed to provide forest users with some voice in policy formulation and implementation at both local and national levels. We can facilitate this involvement in several ways.

Firstly, platforms need to be created for people to meet and for ideas to be debated publicly. Our experience suggests that forest departments should neither convene nor facilitate discussions, as they tend to defend their own interests in the process. Third parties can better consider the perspectives of government foresters, local forest users and other stakeholders, and facilitate dialogue among them. NGOs have often filled this role; but as cases in India and the Philippines indicate, their independence must be assured. Local administrations empowered under PESA in India provide an important example of a government institution acting as third party. Even with an independent facilitator, the least powerful forest users – women, ethnic minorities, the poor and low caste – will need special protections and support in any multi-stakeholder negotiations over forest management. Legal literacy campaigns in India and the Philippines have helped local users to negotiate more effectively and deserve more attention. We can work with disadvantaged groups to define what other forms of support are needed.

Secondly, we can support popular mobilization over forest issues. Federations and other apex organizations have already formed in India and the Philippines to articulate the needs of forest user groups and to exert political pressure on forest departments and other state agencies. As evidence from both India and the Philippines suggests, however, precautions must be taken to ensure that such organizations represent local forest users fairly. We can help the organizations develop mechanisms to ensure their continued account-ability to their constituents. We can also help them to protect their financial and political autonomy, something that has been a problem in our case sites. Where local political practice excludes women and other marginalized groups, we can encourage the development of separate organizations that overlap with existing apex organizations. Where apex organizations are not yet common, we can support NGOs as temporary surrogates, a role they have played in the Philippines and India. These should not be a permanent substitute for people's organizations.

## Create clearer and fuller property rights at the local level

Our research consistently pointed to weak forest and forest product tenure as an important reason that local forest users had not benefited from devolution policies. Without legally recognized property rights, local people's control over forest management was subject to re-interpretation and appropriation by foresters and others. In China, locals enjoyed greater tenure rights under household management systems than anywhere else in our research. Though the harvesting and sale of timber were still heavily regulated, tree 'owners' were able to treat their trees as an asset: to manage them, dispose of them and enjoy the products and/or income from that disposal. *Van panchayats* in India exercised similar tenure rights, though as a collective. In most of India and the Philippines, however, tenure rights are diluted and ambiguous. We can support the development of property rights that include clear resource boundaries and rights of management, use, sale and disposal in these areas.

Where the state must protect non-local interests, we can help to identify incentives, regulations and sanctions associated with specific *outcomes* of local management (Bennett, 1998), as opposed to mandating specific management activities. For example, we can help develop mechanisms for downstream communities to pay those upstream for watershed protection when water quality standards are met. We can also help local users and government officials to determine if easements on local forest tenure are an appropriate strategy for preventing wholesale land conversion without further regulating forest management practices. Of course, any qualification of local tenure can become overly restrictive of local livelihood options (Li, 1999) and brings with it the threat of further restrictions in the future. We can avoid this problem by retaining a substantive focus on maintaining healthy and viable forest-based livelihoods and by adhering to procedural norms that provide local forest users, especially disadvantaged groups, rights of contribution and refusal in the design and revision of any agreements to limit their tenure.

Clearly, the state will assume tenure of some forest to ensure adequate timber production and biodiversity conservation and to meet other national priorities. As our cases suggest, however, expanding the protected area network or state-run timber plantations often means drastically reducing the livelihood options of local forest users. It occasionally threatens their very lives. We have also learned that states often claim lands that are ill suited to forestry, or that have been dominated by other land uses for many generations. A negotiation process is needed to determine when forestry claims to land are legitimate and rational and when they are not. This process must include extensive input from local land users whose survival is affected by the decision.

We can encourage multi-stakeholder reviews of any proposal to reserve forests as state property for conservation, timber production and other activities justified in the national interest.

## Support for rural livelihoods must be more central to devolution policies

Most of the devolution policies considered in these cases were justified, at least partially, in terms of their support for local livelihoods. Yet, most of the provisions of the co-management arrangements, leases, reforestation programmes and contracts emphasize restrictions and requirements that limit local forest users' livelihood choices rather than expand them. Technical and management support for products tended to focus on products valued commercially, such as timber and some NTFPs, which may or may not support a large number of local livelihoods. Forest management must give more attention to improving local livelihoods, not just meeting the forest departments' interests in conservation and commercial revenues.

The first step is a reworking of the definition of forest in many countries. Emphasis on timber species, canopy cover or wildness in many policy implementation documents contributes to the invisibility of forest-based livelihoods such as collection, use and sale of poles, fuelwood, medicinal plants, fodder grasses and a wide range of foods, many of which are critical to the well-being of local residents. We can help to promote definitions of good forest that permit these and other uses in addition to (and even at the expense of, in some cases) the production of woody species. We can also help to set standards for *actively identifying the full range of livelihoods* associated with a particular forest before any management plans are agreed to, as well as standards for how to address the potential impacts of forest management on those livelihoods.

Finally, we believe that forest users should be compensated for livelihoods forgone in the name of conservation or other public interest objectives. This is still very rare in our case study countries. The local government in one site in China has considered the issue of compensation, but still needs to develop a process for involving forest users in defining the terms. The process will be a delicate one, as the promise of compensation will almost certainly elicit false and exaggerated claims. We can encourage government, industry, conservation organizations and others to include provisions for negotiating the terms of compensation with those injured by forest management policy and planning, and help to establish guidelines that prevent the abuse of compensation agreements.

## Policy must make provisions for building local capacity

While it is essential to 'start with what is there' in terms of local forest management, our case studies indicate an urgent need for local capacity-building if forest users, especially the poorest among them, are to take advantage of policy innovations and influence their reform. Capacity-building is needed in several areas: technical skills, marketing, organizational development, communication, legal literacy and political mobilization. However, we feel local capacity is especially weak in dealing with local inequities and exploitative social relations, and in addressing inter-village problems and opportunities.

When we are involved in building local capacity, we must acknowledge exploitative social relations and target our efforts on those who need it most – politically, culturally or economically disadvantaged groups. We can encourage the development of separate organizations, require minimum representation of groups in 'community' organizations, and develop effective mechanisms for safeguarding the well-being of weaker groups. We can also encourage government to transfer forest management authority to forest user groups as such, rather than to communities. These and other strategies will help us to address inequitable and exploitative social relations.

Much of the current capacity-building in our case study countries emphasizes small-scale, village-based organizational development. In many places, however, a more pressing need is to promote networking among forest user groups to address inter-village problems and opportunities, especially the absence of mechanisms for resolving inter-village disputes or setting inter-village boundaries for forest management authority. Networks can also help to ensure that forest users have influence over development projects, industrial investments and other potential large-scale threats to their livelihoods and decision-making processes. As our cases indicate, models for networking are many and varied – ranging from occasional farmer-to-farmer visits to federations with legal standing and more or less permanent institutional presence. We can work with forest users to design and implement the most appropriate model for the local circumstances.

## Shift the focus of state and NGO interventions to issues of political process and away from the technical and managerial aspects of forestry

Bebbington (1998b) has pointed out that technical and managerial capacity-building must accompany political mobilization if rural livelihoods are to be sustained. Our experience suggests that until basic issues of political process have been worked out, technical assistance is likely to reinforce the conceptual models and support the interests of foresters at the expense of local users. Too often, specific management goals and activities were determined by the 'capacity-builders' rather than by forest users. Because many NGOs depend upon donors and government agencies for their funding, legal status and professional opportunities, their assistance often did not meet the needs of forest users much better than that of the forest department (Alsop, et al, 2000; Jenkins and Goetz, 1999; Sen, 1999; Bebbington, 1998a).

We believe that, for the time being, capacity-building in our case study countries should focus on procedural questions of representation, accountability and transparency within local organizations and their networks. Assistance should allow for diverse constitutional forms to exist provided that certain democratic standards are met. We can help to build links between forest user groups and social development and empowerment programmes initiated outside of forest departments, such as the *Mahila Samakhya* Programme for women's empowerment in India, that have already made progress in establishing democratic decision-making procedures. We can also help to ensure that local government is accountable to forest users. Many local governments now have access to expanded financial resources, as well as obligations to provide local services. In China and the Philippines, this often meant that local users and local governments competed for shares of income from timber and other forest products. However, in parts of Madhya Pradesh where PESA was taken seriously, local governments have been more responsive to the demands of forest users, even the poorest among them.

Once some form of democratic decision-making is in place, forest users can decide what technical and managerial assistance is necessary and how to get it. In some places, forest users have attempted to exercise control over substantive issues by contracting foresters and other experts to provide technical and organizational assistance (see Chappela, 2000; Alsop et al, 2000). Contracts for assistance are likely to work best if forest users exert substantial control over the terms of reference of the capacity-builders and if relatively independent financing for the assistance can be arranged. Trust funds for user groups may give groups more control over how support is provided to them. We can help forest users to explore trusts as a means of ensuring some measure of local independence in, and control over, how technical and managerial assistance is obtained.

Devolution policies continue to spread in our case study countries. While they sometimes offer modest advantages to local forest users, their limitations are now quite evident. A new generation of policies, developed and implemented in the spirit of pluralism and democratic accountability, is needed. We can help to bring this generation into being by taking direction from the ongoing efforts of forest users to change the way that forests are managed. The alternative will be to watch the pressure on forests and rural livelihoods mount under the current set of policies.

# Notes

## CHAPTER 1

1   See Agrawal and Ostrom (2001); Enters et al (2000); Agrawal and Ribot (2000); Manor (1999); and Rondinelli (1981) for a discussion of the diverse motivations associated with support for devolution policies.
2   For a discussion of the history of forest management in Asia, see the special issue of *Environmental History* (2001) 6 (2), April.
3   We use the term 'state' to denote not only central government bureaucracies and political institutions, but also their relationship with powerful interests that influence government decision-making, especially donors and private corporations. To assume that official policies represent either an uncontested 'public good' or the narrow self-interest of bureaucrats is misleading (Migdal, 1994; Hirschmann, 1999). Instead, they represent a confluence of political forces that often excludes the poorest forest users and their representatives.
4   We use the term 'community' as shorthand for people living in or near forests that often have some level of economic or cultural connection with, and dependence on, those forests and regular contact with those forests. We realize, however, that communities harbour important differences in perspective and capacity to act (see Li, 1996, and Fortmann and Roe, 1993, for a discussion of the uses of the term community).
5   Others frame their discussions in terms of the development of social movements, or the capacity for self-reflexive and historically conscious 'grassroots politics aimed at the creation and defence of spaces for social autonomy' (Veber, 1998, p390; see also Touraine, 1985; Melucci, 1996; Benford, 1997). Bob Fisher has argued that activism on behalf of community forestry in Asia is a burgeoning social movement (pers comm), and analyses of the long history of mass mobilizations of forest users in India (Shiva, 1988; Sundar, 1997; Guha, 2001) would seem to support this claim.
6   According to Dev Nathan, the 1865 Forest Act of India does not provide forest dwellers with the normal protections from state intrusions into their livelihoods – granting foresters the right to make arrests without warrants and to levy fines without due process. The influence of the Indian Forest Department has encouraged other countries to repeat this pattern. The Nepal Forest Act of 1993 also grants foresters rights to make arrests without warrants and to levy fines up to 10,000 rupees.
7   The composition of these teams is described further in each of the country chapters.
8   The project also produced a set of publications related to thematic areas of enquiry in support of the field projects. See special issues in *Environmental History* (2001) 6 (2); the *International Journal of Agricultural Resources, Governance and Ecology* (2001) 1 (3/4), as well as a special CIFOR publication, Wollenberg (2001), from a workshop held at the East–West Center entitled 'Sharing Innovations: Methods for Multiple Stakeholder Management of Community Forests', Honolulu, August, 1999.

9    See Wollenberg and Ingles (1998) for discussions of dependence on non-timber forest products (NTFPs). However, see CIFOR publication (Arnold and Ruíz-Perez, 1998) on the applicability of the term forest dependence.

10   See Harriss and De Renzio (1997) for a general discussion of social capital.

11   See Mayers and Bass (1999); Manor (1999); and Migdal (1994) for a discussion of how policies contradict one another, weaken and metamorphose over time.

12   See Gilmore and Fisher (1997) and Malla (2001) for a discussion of the influence of external actors on forest policy in Nepal.

13   Clearly, this dimension of devolution policy must be understood on a site-by-site basis as local practices vary widely from site to site, even if policies are uniform (and policy uniformity is probably exaggerated, given the scope for those implementing it to make significant changes in how it works).

14   We make claims about the conceptual frameworks of disadvantaged groups with caution. Though the data we collected were weighted towards discussion with poorer forest users, and we tried to involve local forest users in conducting the research in the Philippines and India, we clearly don't speak for them.

15   Analysis of gender differences in China was limited.

16   See Borrini-Feyerabend (1997) for a discussion of community involvement in monitoring. See also Ritchie et al (2000) for a discussion of community-based criteria and indicators of forest quality, and Blauert and Zadek (1998) for a more general discussion of the positive potential of indicators.

17   We recognize that these categories have permeable boundaries. The evaluation of environmental services, for example, touches upon issues of forest quality and livelihoods. We have developed the categories as a means of ensuring that basic points are covered in the research, not as a rigid framework for the analysis of our findings.

# CHAPTER 2

1    The (administrative) village is basically the equivalent of a production brigade (see also endnote 2). It is now governed by a village committee that is defined, by law, as a self-governing body rather than as a level of government (see Shi, 1999). In practice, however, it remains largely a local arm of the government. Given this ambiguous status, we refer to administrative villages in this chapter neither as 'civil society', nor as the 'lowest level of government'. Instead, the term 'local entity' is used.

2    The commune had a three-tier structure: the commune, production brigade and production team, in descending geographic and administrative scale. After the formal demise of the commune in 1984, the administrative hierarchy was reconstituted as the township, (administrative) village and villager household group (usually all households in a hamlet – *cunminzu* in Chinese), respectively.

3    Land distribution and land leasing in this chapter refer to use rights to land, rather than land ownership. In China, land is owned by either the collective or the state and cannot be transferred.

4    There are many terms with respect to the transfer of land in China: *churang* (transfer), *chengbao* (contract), *zulin* (lease), *lianying* (co-operative undertaking) and *paimai* (auction). Different people use the same term for different things, while different terms are used for the same thing. In this chapter, land lease refers to obtaining use rights to land through paying rent fees in one payment or by installment. Actions to obtain land-use rights by sharing profits at the point of harvest or sale are grouped as a broad category (self-initiated shareholding systems).

# CHAPTER 3

1   Areas with tribal majority populations to which special provisions for safe-
    guarding tribal identity, culture and interests are applicable under Schedule V of
    the Indian constitution.

2   Research work in Orissa was coordinated by Neera Singh, then secretary of
    Vasundhara, a non-governmental organization (NGO) actively involved with
    community forest management (CFM) issues in Orissa. In MP, the research work
    was undertaken by Dr Nandini Sundar, then reader at the Institute of Economic
    Growth, Delhi, and Ranu Bhogal. The Uttarakhand study was undertaken by
    Madhu Sarin, with fieldwork for village case studies by Tarun Joshi, Geeta Gairola
    and Seema Bhatt. The different state studies were jointly coordinated by Madhu
    Sarin and Neera Singh.

3   In November 2000, three new Indian states were created by sub-dividing UP, MP
    and Bihar. Bastar now falls in the new state of Chhattisgarh, carved out of MP,
    and Uttarakhand falls in the new state of Uttaranchal, carved out of UP. As the
    research predated creation of the new states, this chapter refers to them as falling
    in MP and UP, respectively.

4   According to Planning Commission of India estimates, between 1951 and 1990
    alone, 21.3 million persons were displaced for various 'development' projects.
    8.54 million (40 per cent) of the displaced persons belonged to scheduled tribes
    constituting only 8 per cent of the total population inhabiting forested regions.
    Only 2.1 million of the tribals are reported to have been rehabilitated and the
    remaining 6.4 million were left to fend for themselves (Bhuria, 2001).

5   The term *adivasi* literally means indigenous people and is commonly used to refer
    to the tribal population.

6   Most state JFM orders require the preparation of local need-based micro-plans.
    State forest departments, however, also have 10- to 20-year legally binding work-
    ing plans specifying technical management prescriptions. The enforceability of
    JFM micro-plans not conforming to working plan prescriptions remains clouded
    in uncertainty.

7   The study of federations built upon an earlier study of six federations of CFM
    groups from a gender and equity perspective, undertaken by Vasundhara (unpub-
    lished) with the Department for International Development (DFID) funding.

8   Pers comm with Manoj Patnaik of the NGO Regional Centre for Development
    Communication (RCDC), April, 2001. This is based on extrapolation from an
    almost completed survey of CFM in all districts of Orissa.

9   Lower castes are often socially excluded within communities; but village-based
    norms exist for observing festivals together.

10  In late 2002, US$1 was approximately equal to 49 rupees; in 1989, US$1 was
    equivalent to about 30 rupees.

11  In 1942, a draft 'Forest Policy for the Eastern Feudatory States' (under the Eastern
    States Agency) was circulated to obtain comments from various princely rulers
    in Orissa. This policy stated that the first claim on any forests is that of the local
    population. Village forests were to be managed by village *panchayats*. Although
    most rulers approved the draft policy, no systems for operationalizing it were
    established.

12  In addition to the village case studies, this section has drawn on community
    perceptions articulated at various meetings and workshops in Orissa over the past
    seven years. During 1998 to 1999, overlapping with this research, Vasundhara and

Sanhati (an NGO alliance) were also involved in facilitating broad-based discussion and debate on a pro-people forest policy with CFM groups all over Orissa, supported by Action-Aid (India). Nineteen district workshops, several local meetings in different districts and two state-level workshops were organized for this.

13   According to recent forest department data, the number of *vana samrakhan samitis* in the state had increased to 5931 by August 2001 (OFD, 2001). It is unclear how the number of official *samitis* had increased so rapidly. Many of the *samitis* initiated by the department exist only on paper, formed in a rush to show results.

14   In 1991, the per capita forest area in Bastar was 9.5 hectares, compared to 0.30 hectares for Madhya Pradesh and 0.11 hectares for the whole of India (GoMP, 1998, p208).

15   Note prepared by Divisional Forest Officer, Working Plans, Kanker.

16   In return for use of the forest, the other villages each had to bring 100 rupees, a goat and 15 to 30 kilograms of rice to the Ulnar *dharni jatra* (festival at the shrine of the earth) in December and to the *chornia mandai* in April.

17   The practice of giving watchmen a khaki uniform and the *jungle sarpanches* a turban apparently stopped in the mid 1990s.

18   The last two orders were issued after fieldwork for this study was completed.

19   Out of the almost 12,000 VFCs and FPCs constituted, only 1140 received generous funding under the World Bank Forestry Project. The rest received limited or no funding, except where funds from other government programmes could be mobilized.

20   Phase I of this project ended in 1999. A proposed phase II is now being designed after being held up until mid 2002 due to allegations by mass tribal organizations that phase I had violated the land, forest and human rights of poor tribal people.

21   Despite administration orders prohibiting assemblies, about 2500 tribals assembled outside of the Harda district collector's office on 24 December 2001 to demand details about their 'community' accounts and an end to other forms of exploitation by forestry staff (pers comm with *Shramik Adivasi Sanghatan*, 1 January 2002).

22   Literally, setting the reserve forests ablaze in protest against denial of customary access to them.

23   The term used for civil lands in the adjoining state of Tehri Garhwal.

24   In 1986, the ban was made applicable above an altitude of 2500 metres. At lower altitudes, green felling of only pine in areas specified in forest working plans is permitted (Saxena, 1995b).

25   Forest area constitutes 67 per cent of Uttarakhand's total geographical area. Whereas the reserve forests are exclusively under the forest department's jurisdiction, in the case of civil/*soyam* and *van panchayat* forests, the forest department is responsible only for technical supervision. The relative area under civil/*soyam* and *panchayati* forests changes with the conversion of more of the former into the latter.

26   Interestingly, although the Scheduled Districts Act was repealed in 1935, the 1931 rules that were framed under it continued to exist until 1976, when they were revised under section 28 of the Indian Forest Act. Rights activists in Uttarakhand feel that this was illegal and could still be challenged.

27   In Uttarakand, the head of the *van panchayat* is called *sarpanch*, whereas the head of the *gram panchayat* (the smallest unit of local government) is called a *pradhan*. In most other states, the head of the *gram panchayat* is called *sarpanch*.

28   Although leasing, even for non-forestry purposes, was, and still is, permitted with the permission of the deputy commissioner.

29   This included timber.
30   According to an often quoted estimate (Saxena, 1995b; Singh, 1997a) the existing tree cover on reserve, civil/*soyam* and *panchayat* forest lands is 50 per cent, 10 per cent and 40 per cent of their potential, respectively. Civil and *soyam* lands are more heavily degraded because they are treated as open-access forests. In contrast, *panchayat* and reserve forests are protected and managed by local communities and the forest department, respectively (Saxena, 1995b). Reserve and civil forests, however, have benefited from a huge advantage in funding compared to *van panchayats*.
31   One *lakh* is equivalent to 100,000.
32   According to a study of *van panchayats* in the Gairsain block of Chamoli District, one *gram panchayat* has been split into 20 *gram panchayats* and 32 *van panchayats* since the late 1950s (PSS, 1998).
33   Until the time of our research, VFJM mainly covered existing *van panchayat* forests and civil/*soyam* lands, although some reserve forest areas had also been included. No rules, however, had been framed for demarcating reserve forests as village forests. According to recent forest department data, a total of 1217 village forest committees have been constituted for VFJM under a World Bank-supported project. Of these, 729 were VFJM with *van panchayats* and 488 had other village forest committees. Distribution of the forest area brought under VFJM was as follows: *van panchayat* forests, 780 square kilometres; civil/*soyam* forests, 604 square kilometres; and reserve forests, 679 square kilometres (Pai, 2001).
34   The 1997 VFJM rules were revised in late 2001. It is beyond the scope of this chapter to discuss their provisions; but they appear likely to compound existing confusion instead of addressing underlying issues and problems.
35   Emphasis added in this, as well as subsequent clauses of the rules in Box 3.8.
36   Intriguingly, the forest department of the new state of Uttaranchal, under which Uttarakhand now falls, includes all the existing *van panchayats* under its 'achievements' of JFM in the state (see Pai, 2001).
37   The woman president of Khatu Khal Village Forest Committee (by virtue of being the *gram panchayat pradhan*) lived in a village 1.5 kilometres away. The committee accounts were being managed by her son, and the forest guard and the village residents had very little information about them.
38   In the case of Pakhi, an all-women *van panchayat* council was subsequently elected. However, the woman *sarpanch's* husband was the forest guard, muting her questioning voice (see Sarin, 2001b).

# CHAPTER 5

1   We thank Bob Fisher for his help in developing these themes and organizing their presentation.
2   The rewards and sanctions that Said (1978) discusses help to explain the divergent interests among those making claims on the forest, and suggest that 'retraining' foresters and other state officials may not be sufficient to change their daily practice. Incentives, pressures and new mechanisms of accountability may be necessary. We share Bevir's (1999) belief, however, that the habits can be broken, and that power differences do not imply an inevitable, complete or permanent domination of the poorest forest users.

3   The state's emphasis on timber production should not obscure its interest in marketable non-timber forest products (NTFPs), nor the interest among local forest users in obtaining income from such NTFPs. In addition to the case material in Chapter 3, see Olson and Helles (1997), Arnold and Ruíz-Perez (1998) and Yeh (2000) for discussions of tensions and cooperation between local forest users and government officials over the management and use of NTFPs.

4   Some would argue that the state has had a legitimate 'coercive' role in these matters to achieve desirable environmental and economic outcomes (diZerega 2000). We suggest that even where the state has a responsibility to promote conservation or similar 'public interest' objectives, the extent, location and strategy for achieving those objectives should be more heavily influenced by local perspectives and aims of self-determination than it is presently. Recent history in each of our case study countries has seen too much destruction of local livelihoods in the name of public interests, and this imbalance needs to be redressed. Moreover, the fact that each state has accommodated other interests seeking to exploit resources on forest lands, such as logging and mining companies or large dam projects, suggests that the moral high ground of public interest may be a bit lower and less firm than is claimed.

5   The 'theatrical' quality of this rhetoric on the part of government officials should not be ignored, as it speaks of the way that state interests, and not simply old habits of conducting forestry, shape the gap in state and forest-user perspectives (Manor, 1999; see also Alsop et al, 2000).

6   See Scott (1998) for a discussion of 'scientific' forest management's development in Europe, and Guha (2001) for a discussion of how these ideas were incorporated within Indian forestry.

7   By associated companies, we mean both state-run enterprises, common in China, and private companies connected to the state through personal networks, shared personnel, etc.

8   Protests, interventions by advocacy groups and other activities have also increased visibility and legitimacy. In fact, these often gave rise to devolution policies in the first place (see *Environmental History*, 2001). Nevertheless, policies have kept the momentum going where actions were already underway, and created a political platform for those who had not yet made their voices.

9   See Melucci (1996) for a general discussion of the problems of 'visibility' for oppositional groups.

10  Even in areas where the forest department's management has patently failed earlier, as in much of the reserve and protected forests of India and throughout the Philippines.

11  Fiscal decentralization in China and the Philippines has encouraged local governments to turn to forest product taxes and fees as an important source of revenue, particularly in isolated areas with few other valuable commercial products.

12  See Lindsay (1998) for a discussion of various legal mechanisms for protecting local interests, including various contractual arrangements.

13  See Ribot (1999) for a discussion of accountability measures that might be instituted.

14  In most of our case study sites, local institutions are community-based organizations, though households have become forest managers in many parts of China, and local governments have taken on new management responsibilities in all of our case countries.

15  Sivaramakrishnan (2000) discusses 'working plans as instruments of remote control' at some length. Ironically, efforts in the Philippines and India to make planning more participatory through rapid appraisal exercises and locally relevant through micro-planning have served more to align local interests with those of forest agencies, rather than the reverse. In the Philippines, for example, community-based forest management plans and resource-use plans had to be approved by the Department of Natural Resources. In India, micro-plans needed to be approved by district forestry officials. In China, by way of contrast, village leaders could approve plans on collective forests, and households did not need to submit any plans on family plots.

# CHAPTER 6

1  But not necessarily all of the people within villages.
2  See Chappela (2000) and Klooster (2000) for examples from Mexico.
3  See also Wiley (1997) and Massawe (2001) for examples from Tanzania.
4  As noted earlier, we use the term 'state' to refer to governments and the various private, often corporate, interests, donors and NGOs that work closely with them. We therefore use the term government here to note where officials may act against corporate interests that they normally support.
5  We leave aside the issue of individual or clandestine resistance, though these certainly represent an important assertion of rights to make decisions about forests (Scott, 1976; Peluso, 1992; Klooster, 2000).
6  The term outsider may not be appropriate in some cases, as local government representatives or NGO staff may well come from, or reside in, an area where forests are used directly for subsistence and market activities. Moreover, some people who are heavily dependent upon the forest for their livelihoods, and with social and cultural networks woven through a particular forest, may actually reside outside (see Li, 1996; Enters and Anderson, 1999).

# References

Agarwal, C (1996) 'Boundary and Property Rights in Uttarakhand Forests', *Wastelands News*, New Delhi, February–April: 4–6

Agrawal, A and C Gibson (1999) 'Enchantment and Disenchantment: the Role of Community in Natural Resource Conservation', *World Development*, 27: 629–649

Agrawal, A and E Ostrom (2001) 'Collective Action, Property Rights, and Decentralization in Resource Use in India and Nepal', *Politics and Society*, 29(4): 485–514

Agrawal, A and J Ribot (2000) *Decentralization, Participation and Accountability in Sahelian Forestry: Legal Instruments of Political-administrative Control.* Center for Population and Development Studies, Harvard University, Harvard

Albers, H, S Rozelle and G Liu (1998) 'China's Forests Under Economic Reform: Timber Supplies, Environmental Protection, and Rural Resource Access', *Contemporary Economic Policy*, 16: 22–33

Alsop, R, E Gilbert, J Farrington and R Khandelwal (2000) *Coalitions of Interest: Partnerships for Processes of Agricultural Change.* Overseas Development Institute, London

Anderson, J, J Clément and L V Crowder (1999) 'Pluralism in Sustainable Forestry and Rural Development: an Overview of Concepts, Approaches and Future Steps' in *Pluralism and Sustainable Forestry and Rural Development. Proceedings of the International Workshop on Pluralism and Sustainable Forestry and Rural Development.* FAO, Rome, 9–12 December 9-12 1997

Arnold, J E M and M Ruíz-Perez (1998) 'The Role of Non-timber Forest Products in Conservation and Development' in E Wollenberg and A Ingles (eds) *Incomes From the Forest: Methods for the Development and Conservation of Forest Products for Local Communities.* Center for International Forestry Research, Bogor, Indonesia, pp17–42

Arnold, M (1989) 'Ten Years in Review', *Community Forestry Note 7.* FAO, Rome

Baland, J and J Platteau (1996) *Halting Degradation of Natural Resources: Is There a Role for Rural Communities*, Clareden Press, Oxford

Baviskar, A (2001) 'Forest Management as Political Practice: Indian Experiences with the Accommodation of Multiple Interests', *International Journal of Agricultural Resources, Governance and Ecology*, 1(3/4): 243–263

Bebbington, A (1997) 'Social Capital and Rural Intensification: Local Organizations and Islands of Sustainability in the Rural Andes', *The Geographical Journal*, 163, 2: 189–197

Bennett, C (1998) 'Outcome-based Policies for Sustainable Logging in Community Forests: Reducing Forest Bureaucracy' in E Wollenberg and A Ingles (eds) *Incomes from the Forest*, Center for International Forestry Research, Bogor, pp203–220

Bourdieu, P (1986) 'The Forms of Capital' in J G Richardson (ed) *Handbook of Theory and Research for the Sociology of Education*, Greenwood, New York

Bebbington, A (1998a) 'NGOs: Mediators of Sustainability/Intermediaries in Transition?' in J Blauert and S Zadek (eds) *Mediating Sustainability: Growing Policy from the Grassroots*. Kumerian Press, West Hartford, CT

Bebbington, A (1998b) 'Sustaining the Andes? Social Capital and Policies for Rural Regeneration in Bolivia', *Mountain Research and Development* 18(2): 173–181

Behar, A and R Bhogal (2000) 'People's Social Movements: An Alternative Perspective of Forest Management', Paper produced under the CIFOR Research Project Creating Space for Local Forest Management, New Delhi

Benford, R (1997) 'An Insider's Critique of the Social Movement Framing Perspective', *Sociological Inquiry*, 67(4): 409–430

Bevir, M (1999) 'Foucault, Power, and Institutions', *Political Studies*, 47, 2: 345–359

Bhatt, S (1999) 'A case study of some people's institutions in the Akash Kamini Valley, Garhwal', unpublished paper for the CIFOR research project on Creating Space for Local Forest Management, New Delhi

Bhogal, R K and T S Bhogal (2000) 'Joint Forest Management in Harda', Draft paper under the CIFOR Research Project Creating Space for Local Forest Management, New Delhi

Bhogal, R K and M Shankar (2000a) 'Nationalised Forest Produce: A Study of the *Tendu Patta* Policy of Madhya Pradesh, India. A Devolution or a Welfare Policy?' Paper under the CIFOR Research Project Creating Space for Local Forest Management, New Delhi

Bhogal, R K and M Shankar (2000b) 'The Van Dhan Initiative in Central Bastar', Paper produced under the CIFOR Research Project on Creating Space for Local Forest Management, New Delhi

Bhuria, D S (2001) Address given by Dilip Singh Bhuria, Chairman, National Commission for Scheduled Castes and Scheduled Tribes, Government of India, at the National Meeting on Community Forestry Issues, organised by the World Bank, Manesar, Haryana, 8–9 November 2001

Blauert, J and S Zadek (1998) *Mediating Sustainability: Growing Policy from the Grassroots*, Kumarian Press, West Hartford, Connecticut

Boaz, A A (1998) *Management of Non-nationalized Minor Forest Produce of Madhya Pradesh*. A case study of the Madhya Pradesh Minor Forest Produce Federation, Bhopal

Borrini-Feyerabend, G (ed) (1997) *Beyond Fences: Seeking Social Sustainability in Conservation*, volumes 1 and 2. IUCN, Gland, Switzerland

Borrini-Feyerabend, G, F M Taghi, Ndangang and J C Nguinguiri (2000) 'Co-management of Natural Resources', IUCN and GTZ, Germany

Bray, D (2000) 'Adaptive Management, Organizations and Common Property Management: Perspectives from the Community Forests of Quintana Roo, Mexico', Paper presented at the Eighth Biennial Conference of the International Association for the Study of Common Property, 31 May–4 June, Bloomington, IN

Brown, K and S Rosendo (2000) 'Environmentalists, Rubber Tappers and Empowerment: The Politics and Economics of Extractive Reserves', *Development and Change*, 31: 201–228

Bruce, J W et al (1995) 'Experimenting with Approaches to Common Property Forestry in China', *Unasylva*, 46 (180): 44–48

Campbell, J (1992) 'Changing Objectives, New Products and Management Challenges: Making the Shift from 'Major' vs 'Minor' to Many Forest Products, Paper presented at the Seminar on Forest Products, Coimbatore, India

CECI (Canadian Centre for International Studies and Cooperation) (1998) 'Community-based Economic Development Project, India: Preliminary Proposal', mimeo, CECI, Montreal

Chambers, R (1992) 'Participatory Rural Appraisals: Past, Present and Future', *Forest Trees and People Newsletter*, 15/16: 4–9

Chandrasekharan, D (1997) *Forest, Trees And People. Conflict Management Series.* Proceedings of the Electronic Conference on Addressing Natural Resource Conflicts through Community Forestry, January–May 1996, Community Forestry Unit, FAO, Rome

Chappela, F (2000) 'The Development of Community Forestry in Mexico', Unpublished manuscript under the CIFOR Research Project Creating Space for Local Forest Management, New Delhi

Chen, F and Q Gao (1997) 'Viewpoint Shared by Studies of Stocked Cooperatives Economies in China', *Forestry Economy*, 1: 65–68

Chiong-Javier, E (1996) 'Hanging Out on a Limb: the ISFP Devolution Experience' in J Marco and E Nuñez, Jr (eds) *Participatory and Community-based Approaches in Upland Development: A Decade of Experience and a Look at the Future.* Philippine Uplands Resource Centre, the Philippines

Chiong-Javier, M A and J T Dizon (2003) 'Creating Space in Obo, Dalaguete, Cebu' in A P Contreras (ed) *Creating Space for Local Forest Management in the Philippines*, La Salle Institute of Governance (LSIG) and Antonio Contreras, Manila, Philippines, pp105–120

Colchester, M (1994) 'Sustaining the Forests: The Community-based Approach in South and South-East Asia', *Development and Change*, 25(1): 69–100

Coleman, J S (1990) *Foundations of Social Theory*, Belknap Press, Harvard University Press, Cambridge

Conroy, C, A Mishra and A Rai (2000) 'Learning from Self-initiated Community Forest Management in Orissa, India', *Forests, Trees and People Newsletter* No 42: 51–56

Contreras, A P (ed) (2003) *Creating Space for Local Forest Management in the Philippines*, La Salle Institute of Governance (LSIG) and Antonio Contreras, Manila, Philippines

Corbridge, S and S Jewitt (1997) 'From Forest Struggles to Forest Citizens? Joint Forest Management in the Unquiet Woods of India's Jharkand', *Environment and Planning A*, 29: 2145–64

Daniels, S and G Walker (1999) 'Rethinking Public Participation in Natural Resource Management: Concepts from Pluralism and Five Emerging Approaches' in *Pluralism and Sustainable Forestry and Rural Development.* Proceedings of an International Workshop, FAO, Rome, 9–12 December 1997

Daowei Z (2001) 'Policy Reform and Investment in Forestry in China', Paper presented at the International Symposium Chinese Forest Policy, held in Sichuan Province, China, June 2001

Das, V (2002) 'Struggle for Land and Forest Rights' in *Women's Empowerment Policy and Natural Resource – What Progress?* Proceedings of a workshop jointly organized by the Planning Commission, GOI, and the Overseas Development Group, University of East Anglia, UK, Write-Arm, Bangalore

del Castillo, R and S Borlagdan (1996) 'Participatory Initiatives in Upland Development: Over a Decade of Institution Building' in J Marco and E Nuñez, Jr (eds) *Participatory and Community-based Approaches in Upland Development: A Decade of Experience and a Look at the Future.* Philippine Uplands Resource Center, the Philippines

Dermawan, A and I A P Resosudarmo (2002) 'Forests and Regional Autonomy: The Challenge of Sharing the Profits and Pains' in C J P Colfer and I A P Resosudarmo (eds) *Which Way Forward? People, Forests and Policymaking in Indonesia*, Resources for the Future, Washington, DC, pp325–357

Diwan, R, M Sarin and N Sundar (2001) *Jan Sunwai* (Public Hearing) on Forest Rights at Village Indpura, Harda District, 26 May 2001, mimeo (later published in *Wastelands News*, May–July 2001, XVI(4) and in *Van Sahyog*, May–July, 2001, 3(1))

diZerega, G (2000) *Persuasion, Power and Polity: A Theory of Democratic Self-Organization*, Hampton Press, Cresskill, New Jersey

Dizon, J T and J M Servitillo (2003) 'Creating Space in Sangbay, Nagtipunan, Quirino' in A P Contreras (ed) *Creating Space for Local Forest Management in the Philippines*, La Salle Institute of Governance (LSIG) and Antonio Contreras, Manila, Philippines, pp39–52

Doornbos, M, A Saith and B White (2000) 'Forest Lives and Struggles: An Introduction', *Development and Change*, 31(1): 1–10

*Down to Earth* (2000) 'Community Forest Management: The Nepalese Experience', *Down to Earth*, 8(19): 30–46

D'Silva E and Nagnath, B (2002) 'Behroonguda: A Rare Success Story in Joint Forest Management', *Economic and Political Weekly*, 9 February: 551–557

Dubey, A P (Additional Principal Chief Conservator of Forests–JFM, MP) (2001) District office No 697, 6 July 2001, to MOEF in response to its query about the Harda Public Hearing, circulated on the DNRM and RUPFOR email discussion groups

Dubey, R, Y K Pant and N S Jeena (2000) *Van Panchayat, Ek Parichay*. Janpad (District Administration), Nainital, Uttar Pradesh

Dubois, O (1998) 'Capacity to Manage Role Changes in Forestry: Introducing the "4Rs" Framework', *Forest Participation Series, No 11*. International Institute for Environment and Development, London

Duhaylungsod, L A, V Buhisan and J Duhaylungsod (2003) 'Creating Space in Up, Maitum, Sarangani' in A P Contreras (ed) *Creating Space for Local Forest Management in the Philippines*, La Salle Institute of Governance (LSIG) and Antonio Contreras, Manila, Philippines, pp199–210

Ecotech Services (1999) *Study on Management of Community Funds and Local Institutions, State Perspectives and Case Studies*, volume II, draft for discussion, mimeo, New Delhi

Edmunds, D and E Wollenberg (2001) 'A Strategic Approach to Multistakeholder Negotiations, *Development and Change*, 32(2): 231–253

Engel, P G H, A Hoeberichts and L Umans (2001) 'Accommodating Multiple Interests in Local Forest Management: A Focus on Facilitation, Actors and Practices', *International Journal of Agriculture, Resources, Governance and Ecology* (IJARGE), Special issue on accommodating multiple interests in local forest management, 1(3/4): 306–326

Enters, T, and J Anderson (1999) 'Rethinking the Decentralization and Devolution of Biodiversity Conservation', *Unasylva*, 50: 6–10

Enters, T, P B Durst and M Victor (eds) (2000) *Decentralization and Devolution of Forest Management in Asia and the Pacific*. RECOFTC Report No 18 and RAP Publication 2000/1, Bangkok, Thailand

*Environmental History* (2001) 6(2), April

Evans, P (1996) 'Government Action, Social Capital and Development: Reviewing the Evidence on Synergy', *World Development*, 24(6): 1119–1132

FDUP (Forest Department of Uttar Pradesh) (1997) *Village Forest Joint Management Rules*. FDUP, 30 August 1997, Lucknow

FDUP (undated a; circa 2000) 'Empowerment of People through Forestry', A status paper on JFM in Uttar Pradesh, Lucknow

FDUP (undated b; circa 2000). *Government Resolution for Eco-Development in Uttar Pradesh*. FDUP, Lucknow

Fisher, R (1999) 'Devolution and Decentralization of Forest Management in Asia and the Pacific', *Unasylva*, 50(4): 3–5

Fortmann, L and J W Bruce (eds) (1988) *Whose Trees? Proprietary Dimensions of Forestry.* Westview Press, Boulder, CO

Fortmann, L and E Roe (1993) 'On Really Existing Communities – Organic or Otherwise', *Telos*, 95: 139–146

Fox, J, L Fisher and C Cook (eds) (1997) *Conflict and Collaboration: Eighth Workshop on Community Management of Forest Lands*, 18 February–28 March 1997. Programme on Environment, East–West Center, University of Hawaii, Honolulu, Hawaii

Froehling, O (1997) 'The Cyberspace "War of Ink and Internet" in Chiapas, Mexico', *The Geographical Review* 87(2): 291–307

FSI (Forest Service of India) (2000) *State of Forest Report 1999.* Forest Survey of India (Ministry of Environment and Forest), Dehradun

Gairola, G (1999a) Field case studies: *CFM in Holta village, district Tehri Garhwal*

Gairola, G (1999b) Field case studies: *Dungri Chopra Van Panchayat, district Pauri Garhwal*

Gairola, G (1999c) Field case studies: *Rejection of VFJM in Naurakh village, district Chamoli*

Gauld, R (2000) 'Maintaining Centralized Control in Community-based Forestry: Policy Construction in the Philippines', *Development and Change*, 31(1): 229–254

Ghildiyal, M C and A Banerjee (1998) 'Status of Participatory Management in Uttarakhand Himalayas', Paper presented at regional workshop on Participatory Forest Management Implications for Policy and Human Resource Development, Kunming, People's Republic of China

Gibson, C, M A McGean and E Ostrom (eds) (2000) *People and Forests: Communities, Institutions and Governance*, MIT Press, Cambridge

Gilmour, D and R Fisher (1998) 'Evolution in Community Forestry: Contesting Forest Resources' in M Victor, C Lang and J Bornemeier (eds) *Community Forestry at the Crossroads: Reflections and Future Directions in the Development of Community Forestry*, Proceedings of an international seminar held in Bangkok, Thailand, 17–19 July 1997, pp27–44

GOI (Government of India) (1988) *National Forest Policy Resolution.* GOI, New Delhi

GOI (1990) *Involvement of Village Communities and Voluntary Agencies in Regeneration of Degraded Forests*, MOEF, No 6.21/89-F.P., 1 June 1990

GOI (1996) *Provisions of the Panchayats (Extension to the Scheduled Areas) Act.* GOI, New Delhi

GOMP (Government of Mahdya Pradesh) (1998) *Madhya Pradesh Human Development Report.* GOMP

GOMP (1999) *Van Dhan Mission: Status Report*, January–September 1999. GOMP, Jagdalpur, Collector's Office

GOO (Government of Orissa) (1996) *Orissa District Gazetteers.* GOO, Cuttack

GOO (2000) *Economic Survey.* GOO, Bhubaneswar

GOU (Government of Uttaranchal) (2001) *The Uttaranchal Panchayati Forest Rules, 2001.* GOU, Dehradun

GOUP (Government of the United Provinces) (1921) *Kumaon Forest Grievances Committee Report.* GOUP, Lucknow

GOUP (1960) *Report of the Kumaun Forests Fact Finding Committee.* GOUP, Lucknow

GOUP (1984) *Van Panchayaton Ki Karya Pranali Ka Moolyankan Adhyayan.* GOUP, Lucknow

Grimble, R J, M-K Chan, J Aglionby and J Quan (1995) 'Trees and Trade-offs: A Stakeholder Approach to Natural Resource Management', *Gatekeeper Series No 52.* International Institute for Environment and Development, London

Grinspoon, E (2001) 'Socialist Wasteland Auctions: Changing Property Rights to Collective Forestland in China's Transitional Economy', Paper presented at International Symposium Chinese Forest Policy, held in Sichuan Province, June 2001

Guha, R (1989) *The Unquiet Woods, Ecological Change and Peasant Resistance in the Himalaya.* Oxford University Press, Delhi

Guha, R (2001) 'The Prehistory of Community Forestry in India', *Environmental History*, 6(2): 213–238

Gupta, A (1995) 'Blurred boundaries: the discourse of corruption, the culture of politics, and the imagined state', *American Ethnologist*, 22(2), 375–402

Gustafsson, B and W Zhong (2000) 'How and Why has Poverty in China Changed? A Study Based on Microdata from 1988 to 1995', *The China Quarterly*, 164: 983–1006

Harriss, J and P de Renzio (1997) '"Missing Link" or Analytically Missing? The Concept of Social Capital', *Journal of International Development*, 9(7): 919–937

Hecht, S and A Cockburn (1989) *The Fate of the Forest: Developers, Destroyers and Defenders of the Amazon.* New York: Verso

Hirschmann, D (1999) 'Development Management Versus Third World Bureaucracies: A Brief History of Conflicting Interests', *Development and Change*, 30: 287–305

Hyde, W (2001) Introductory comments for the International Symposium Chinese Forest Policy, held in Sichuan Province, June 2001

INFORM (2001) Information e-bulletin on Participatory Forest Management, Resource Unit for Participatory Forestry, Winrock International India, vol 1(3), October–December, 2001

*International Journal of Agricultural Resources, Governance and Ecology* (2001) 1(3/4)

Jackson, W J and A W Ingles (1998) *Participatory Techniques for Community Forestry: A Field Manual.* IUCN, Gland, Switzerland, and Cambridge, UK, and World Wide Fund for Nature, Gland, Switzerland

Jena, N S et al (1997) 'Panchayati vano mein vyavsayaik paudharopan', *Parvatiya Vikas*, 1(iv), CDS, Nainital

Jenkins, R and A M Goetz (1999) 'Constraints on Civil Society's Capacity to Curb Corruption: Lessons from the Indian Experience', *IDS Bulletin*, 30(4): 39–49

Johnson, T (1999) 'Community-based Forest Management in the Philippines', *Journal of Forestry*, 97(11): 26–30

Jonsson, S and A Rai (eds) (1994) *Forests, People and Protection, Case Studies of Voluntary Forest Protection by Communities in Orissa.* ISO/Swedforest and SIDA, Bhubaneswar

Joshi, A (1999) 'Progressive Bureaucracy: an Oxymoron? The Case of Joint Forest Management in India', *Rural Development Forestry Network*, Paper 24a, Overseas Development Institute, London

Kafle, G (2001) 'Platforms for Learning: Experiences with Adaptive Learning in Nepal's Community Forestry Programme' in E Wollenberg, D Edmunds, L Buck, J Fox and S Brodt (eds), *Social Learning in Community Forests*, CIFOR, Bogor, Indonesia, pp45–64

Kaimowitz, D, A Faune and R Mendoza (2003) 'Your Biosphere is My Backyard: The Story of Bosawas in Nicaragua', CIFOR Working Paper No 23, CIFOR, Bogor, Indonesia

Kant, S, N Singh, M Singh and K Kundan (1991) *Community Based Forest Management Systems: Case Studies from Orissa.* Indian Institute of Forest Management, Bhopal, and SIDA and ISO/Swedforest, New Delhi

Kekes, J (1993) *The Morality of Pluralism.* Princeton University Press, Princeton, New Jersey

Khare, A K (1993) 'Forest Products Marketing', Madhya Pradesh Integrated Forestry Sector Project Preparation Working Paper 4

Khare, A, M Sarin, N C Saxena, S Palit, S Bathia, F Vania and M Satyanarayana (2000) 'Joint Forest Management: Policy, Practice and Prospects', Paper No 3 in J Mayers and E Morrison (eds) *Policy that Works for Forest and People*. IIED, London

Klooster, D (2000) 'Community Forestry and Tree Theft in Mexico: Resistance or Complicity in Conservation', *Development and Change*, 31(1): 281–306

Lele, S (1998) 'Why, Who and How of Jointness in Joint Forest Management: Theoretical and Empirical Insights from the Western Ghats of Karnataka, Paper presented at the International Workshop on Shared Resource Management in South Asia, Institute for Rural Management, Anand, 17–19 February, Publication No 11, Institute for Social and Economic Change and Pacific Institute for Studies in Development, Environment and Security

Li, T M (1996) 'Images of Community: Discourse and Strategy in Property Relations', *Development and Change*, 27(3): 501–527

Li, T (ed) (1999) *Transforming the Indonesian Uplands: Marginality, Power and Production*. Harwood Academic Publishers, Amsterdam

Lindsay, J M (1998) 'Creating a Legal Framework for Community-based Management: Principles and Dilemmas', Paper presented at the International Workshop on Community-Based Natural Resource Management, 10–14 May 1998, the World Bank, Washington, DC

Liu, D (2001) 'Tenure and Management of Non-state Forests in China Since 1950: A Historical Review', *Environmental History*, 6(2): 239–263

Liu, J (2001) 'Study on the Forestry Taxation and Charges System in South China Collective Forestry Areas', Paper presented at the International Symposium Chinese Forest Policy, held in Sichuan Province, June 2001

Loury, G (1977) 'A Dynamic Theory of Racial Income Differences' in P A Wallace, and A M LaMond (eds) *Women, Minorities and Employment Discrimination*, Lexington Books, Lexington, Mass

Lu W, C Zhang, S Yan, Y Wei, F Yu and X Tan (2001) 'Instruments for Company–Community Partnership in Forestry Sector in China: Focus on APP Case in Guangdong Province', Paper presented at the International Symposium Chinese Forest Policy, held in Sichuan Province, June, 2001

Lynch, O J (1998) 'Law, Pluralism and the Promotion of Sustainable Community-based Forest Management', *Unasylva*, 194: 52–56

Lynch, O (1999) 'Legal Aspects of Pluralism and Community-based Forest Management: Contrasts Between and Lessons Learned from the Philippines and Indonesia' in *Pluralism and Sustainable Forestry and Rural Development*. Proceedings of an International Workshop, FAO, Rome 9–12 December 1997

Lynch, O and K Talbot (1995) *Balancing Acts: Community-based Forest Management and National Law in Asia and the Pacific*. World Resources Institute, Baltimore, MD

Maarleveld, M and C Dangbégnon (1999) 'Managing Natural Resources: A Social Learning Perspective', *Agriculture and Human Values*, 16(3): 267–280

Magno, F (2001) 'Forest Devolution and Social Capital: State–Civil Society Relations in the Philippines', *Environmental History*, 6(2): 264–286

Malla, Y (2001) 'Changing Policies and Persistence of Patron Client Relations in Nepal: Stakeholders' Responses to Changes in Forest Policies', *Environmental History*, (6)2: 287–307

Mandondo, A (2000) 'Situating Zimbabwe's Natural Resource Governance Systems in History', Center for International Forestry Research Occasional Paper No 32. CIFOR, Bogor, Indonesia

Manor, J (1999) *The Political Economy of Democratic Decentralization*. The World Bank, Washington, DC

Marco, J and E Nuñez (eds) (1996) *Participatory and Community-based Approaches in Upland Development: A Decade of Experience and a Look at the Future*, Third National Conference on Research in Uplands, Philippines Uplands Resource Center, Manila

Massawe, E (2001) 'External Donors and Community-based Management of Mgori Forest, Tanzania' in E Wollenberg, D Edmunds, L Buck, J Fox and S Brodt (eds) *Social Learning in Community Forests*. CIFOR and the East–West Center, Bogor, Indonesia

Mayers, J and S Bass (1999) *Policy that Works for Forests and People. Executive Summary of Series Overview*. IIED London

McCarthy, J (2000a) 'The Changing Regime: Forest Property and *Reformasi* in Indonesia', *Development and Change*, 31: 91–130

McCarthy, J (2000b) '"Wild Logging": the Rise and Fall of Logging Networks and Biodiversity Conservation Projects on Sumatra's Rainforest Frontier', Center for International Forestry Research Occasional Paper No 31. CIFOR, Bogor, Indonesia

Melucci, A (1996) *Challenging Codes: Collective Action in the Information Age*, Cambridge University Press, Cambridge

Melucci, A (1996) *Challenging Codes: Collective Action in the Information Age*, Cambridge University Press, Cambridge

Menzies, N and N Peluso (1991) 'Rights of Access to Upland Forest Resources in Southwest China', *Journal of World Forest Resource Management*, 6: 1–20

Migdal, J (1994) 'The State in Society: An Approach to Struggles for Domination' in J Migdal, A Kholi and V Shue (eds) *State Power and Social Forces: Domination and Transformation in the Third World*. Cambridge University Press, New York

Mishra, R (1998) *Legal Framework and Institutional Arrangement between OFD and Communities in Orissa*. Preliminary report of consultant, mimeo, Scandiaconsult Natura AB/Asia Forest Network, Bhubaneswar

Mitra, A (1993) 'Chipko, An Unfinished Mission', *Down to Earth*, 30 April 1993, New Delhi

MOEF (Ministry of Environment and Forests, GOI) (2000) *Guidelines for Strengthening of Joint Forest Management (JFM) Programme*. Forest Protection Division No 22-8/2000-JFM (FPD), 21 February 2000

Moore, M and A Joshi (1999) 'Between Cant and Corporatism: Creating and Enabling Political Environment for the Poor', *IDS Bulletin*, 30(4): 50–59

Morales, B C, M M Paunlagui, and L Paglinawan (2003) 'Creating Space in Maputi, San Isidro, Davao Oriental' in A P Contreras (ed) *Creating Space for Local Forest Management in the Philippines*, La Salle Institute of Governance (LSIG) and Antonio Contreras, Manila, Philippines, pp121–138

MPFD (Madhya Pradesh Forest Department) (undated; circa 2000) *Status of JFM in Madhya Pradesh, Bhopal*. MPFD

MPSMFPCF (2000) *From Gatherer to Owner*. Madhya Pradesh State Madhya Pradesh Forest Department (Trading and Development) Cooperative Federation Ltd, Bhopal

MTOs (Mass Tribal Organizations) (1999) 'Report of the Joint Mission on the Madhya Pradesh Forestry Project', Paper released by the Madhya Pradesh MTOs, Bhopal

Nachmias, D and C Nachmias (1987) *Research Methods in the Social Sciences*. St Martin's Press, New York

Nanda, N (1999) *Forests for Whom? Destruction and Restoration in the UP Himalayas*. Har Anand Publications, New Delhi

Nemarundwe, N (2001) 'Institutional Collaboration and Shared Learning for Forest Management in Chivi District, Zimbabwe' in E Wollenberg, D Edmunds, L Buck, J Fox and S Brodt (eds), *Social Learning in Community Forests*, Center for International Forestry Research, Bogor, Indonesia, pp85–108

O'Brien, K and L Li (2000) 'Accommodating "Democracy" in a One-Party State: Introducing Village Elections in China', *The China Quarterly*, 162: 465–489

Oi, J and S Rozelle (2000) 'Elections and Power: The Locus of Decision-Making in Chinese Villages', *The China Quarterly*, 162: 513–539

Olson, C and F Helles (1997) 'Making the Poorest Poorer: Policies, Laws and Trade in Medicinal Plants in Nepal', *Journal of World Forest Resources Management*, 8: 137–158

Orissa Forest Department (OFD) (1999) *Orissa Forests 1999*. Statistical Branch, Office of the PCCF, Bhubaneswar

OFD (2001). *Status of JFM in Orissa as on 31 August 2001*. Development Circle, Forest Department, Cuttuck

Ostrom, L (1999) 'Self-governance and Forest Resources', *Occasional Paper No 20*. CIFOR, Bogor, Indonesia

Pai, R (2001) 'Community Participation in Management of Natural Resources', Paper presented at authors' workshop for a book on JFM, Delhi, August, 2001

Parajuli, P (1998) 'Beyond Capitalized Nature: Ecological Ethnicity as an Arena of Conflict in the Regime of Globalization', *Ecumene* 5(2): 186–217

Pastor, R and Q Tan (2000) 'The Meaning of China's Village Elections', *The China Quarterly*, 162: 490–512

Pattnaik, S (2001) 'Myth and Reality of Devolution of Power: An examination of PESA in Orissa', Draft paper for the CIFOR Research Project on Creating Space for Local Forest Management, Bhubaneswar

Paunlagui, M M, N J V B Querijero and I V C Ongkiko (2003) 'Creating Space in Paitan, Naujan, Oriental Mindoro' in A P Contreras (ed) *Creating Space for Local Forest Management in the Philippines*, La Salle Institute of Governance (LSIG) and Antonio Contreras, Manila, Philippines, pp165–82

Peluso, N (1992) *Rich Forests, Poor People: Resource Control and Resistance in Java*. University of California Press, Berkeley, CA

Poffenberger, M (ed) (1990) *Keepers of the Forest: Land Management Alternatives in Southeast Asia*. Kumarian Press, West Hartford, CT

Poffenberger, M (ed) (1995) *Communities and Forest Management in Canada and the United States*, IUCN, Geneva, Switzerland

PSS (Panchayat Sewa Samithi) (1998) *People's Forest Management: From Conflict Resolution to Policy Perspectives – An Exploration in the Context of Uttarakhand*. Report submitted to C D S, Nainital, mimeo

PUDR (People's Union for Democratic Rights) (2001) *When People Organise: Forest Struggles and Repression in Dewas*. PUDR, Delhi, June 2001

Pulhin, J M and M L L C Pesimo-Gata (2003) 'Creating Space in Bicol National Park' in A P Contreras (ed) *Creating Space for Local Forest Management in the Philippines*, La Salle Institute of Governance (LSIG) and Antonio Contreras, Manila, Philippines, pp53–66

Putnam, R (1993) 'The Prosperous Community: Social Capital and Public Life', *American Prospect*, 4(13): 35–42

Putnam, R D (1995) 'Bowling Alone: America's Declining Social Capital', *Journal of Democracy*, January: 65–78

Rajan, B K C (1994) *Ten Forest Products*. Jaya Publications, Dehradun

Ramírez, R (2001) 'Understanding the Approaches for Accommodating Multiple Stakeholders' Interests', *International Journal of Agriculture, Resources, Governance and Ecology* (IJARGE), Special issue on accommodating multiple interests in local forest management, 1(3/4): 264–285

Rawat, A S (1998) *Biodiversity Conservation in U P Hills: A People's Viewpoint*, Studies, vol 3(IV). CDS, Uttar Pradesh Academy of Administration, Nainital

Rebugio, L (1996) 'How Participatory and Equitable is the Community Forestry Program: Some Preliminary Evidences' in J Marco and E Nuñez, Jr (eds) *Participatory and Community-based Approaches in Upland Development: A Decade of Experience and a Look at the Future*. Philippine Uplands Resource Centre, the Philippines

Rescher, N (1993) *Pluralism: Against the Demand for Consensus*. Clarendon Press, Oxford, UK

Ribot, J (1998) *Decentralization and Participation in Sahelian Forestry: Legal Instruments of Central Political-administrative Control*, Working Paper Series No 98.06. Harvard Center for Population and Development Studies, Harvard University, Boston

Ribot, J C (2001) 'Integral Local Development: "Accommodating Multiple Interests" through Entrustment and Accountable Representation', *International Journal of Agriculture, Resources, Governance and Ecology* (IJARGE), Special issue on accommodating multiple interests in local forest management, 1(3/4): 327–350

Richards, M (1997) 'Common Property Resource Institutions and Forest Management in Latin America', *Development and Change*, 28: 95–117

Ritchie, B, C McDougall, M Haggith and N Burford de Olivera (2000) *Criteria and Indicators of Sustainability in Community Managed Forest Landscapes: An Introductory Guide*. Center for International Forestry Research, Bogor, Indonesia

Rocheleau, D and L Ross (1995) 'Trees as Tools, Trees as Text: Struggles Over Resources in Zambrana-Chaucey, Dominican Republic', *Antipode* 27(4): 407–428

Röling, N G and M A E Wagemakers (eds) (1998) *Facilitating Sustainable Agriculture: Participatory Learning and Adaptive Management in Times of Environmental Uncertainty*. Cambridge University Press, Cambridge

Rondinelli, D (1981) 'Government Decentralization in Comparative Perspective: Theory and Practice in Developing Countries', *International Review of Administrative Science*, 47(29): 133–145

Rossi, J (1997) 'Participation Run Amok: The Costs of Mass Participation for Deliberative Agency Decisionmaking', *Northwestern University Law Review* 92(1): 173–247

Said, E (1978) *Orientalism*. Vintage Books, New York

Saigal, S (2000) 'Beyond Experimentation: Emerging Issues in the Institutionalization of Joint Forest Management in India', *Environmental Management*, 26(3): 269–281

Saigal, S, C Agarwal and J Campbell (1996) *Sustaining Joint Forest Management: The Role of Non Timber Forest Products*. Society for the Promotion of Wasteland Development, Delhi, mimeo

Salafsky, N, B Cordes, J Parks and C Hochman (1999) *Evaluating Linkages Between Business, the Environment, and Local Communities: Final Analytical Results from the Biodiversity Conservation Network*. Biodiversity Support Program, Washington, DC

Salazar, R (1996) 'Strengthening Participatory Approaches in Upland Development: Communities, NGOs and Local Governments in Focus' in J Marco and E Nuñez, Jr (eds) *Participatory and Community-based Approaches in Upland Development: A Decade of Experience and a Look at the Future*. Philippine Uplands Resource Center, the Philippines

Santos, E P and M A Pollisco-Botengan (2003) 'Creating Space in Presentacion, Camarines Sur' in A P Contreras (ed) *Creating Space for Local Forest Management in the Philippines*, La Salle Institute of Governance (LSIG), Manila, Philippines, pp67–86

Santos, E P with M A P Botengan, E Delloro-Ledesma, M Neonal and L Sucatre, (2000) 'Creating Space for Local Forest Management: The Case of Presentacion Community Forest Management, Camarines Sur, Philippines', Project report.

Sarin, M (1998) 'Who Is Gaining? Who Is Losing? Gender and Equity Concerns in Joint Forest Management', Working paper of the Gender and Equity Sub-group, National Support Group for JFM, Society for Promotion of Wastelands Development, New Delhi

Sarin, M (2001a) 'From Right Holders to "Beneficiaries"? Community Forest Management, Van Panchayats and Village Forest Joint Management in Uttarakhand', Draft paper based on research for the CIFOR project on Creating Space for Local Forest Management, January 2001, Chandigarh

Sarin, M (2001b) 'Disempowerment in the name of "participatory" forestry? Village Forests Joint Management in Uttarakhand', *Forests, Trees and People Newsletter*, No 44, April 2001, Uppsala, Sweden

Sarin, M (2001c) 'Discussion on the Revised Van Panchayat Rules, 2001, and Impacts of Village Forest Joint Management (VFJM) on Van Panchayats in Uttarakhand: Summary of Proceedings', Bhawali, 22 June 2001, mimeo

Sarin, M with L Ray, M S Raju, M Chatterjee, N Banerjee and S Hiremath (1998) *Who Is Gaining? Who Is Losing? Gender and Equity Concerns in Joint Forest Management*, Society for the Promotion of Wasteland Development (SPWD), New Delhi

Sarin M and A Rai (1998) *Rational Distribution of Benefit between Communities and Forest Department, and Within Communities, under JFM and CFM*. Study report prepared for project Capacity Building for Participatory Management of Degraded Forests in Orissa, India, mimeo, Scandiaconsult Natura AB/Asia Forest Network, December 1998, Bhubaneswar

Sato, J (2000) 'People in Between: Conversion and Conservation of Forest Lands in Thailand', *Development and Change*, 31(1): 155–177

Saxena, N C (1995a) *Forests, People and Profit, New Equations for Sustainability*. Centre for Sustainable Development and Natraj, Dehradun

Saxena, N C (1995b) *Towards Sustainable Forestry in the UP Hills*. Centre for Sustainable Development, LBS National Academy of Administration, Mussoorie, Uttar Pradesh

Saxena, N C (1996) 'Forest Policy and Rural Poor in Orissa', *Wastelands News*, XI(2), November–January 1996, New Delhi

Saxena, N C (1997) *You Participate, We Manage: The Saga of Participatory Forest Management in India*. CIFOR Special Publication, Bogor, Indonesia

Saxena, N C (1999) *Forest Policy in India*. WWF-India and IIED, New Delhi

Saxena, N C (2001) *Empowerment of Tribals through Sustainable Natural Resource Management in Western Orissa*. Report prepared for IFAD/DFID, December, 2001

Scott, J (1976) *The Moral Economy of the Peasant*. Yale University Press, New Haven, CT

Scott, J (1998) *Seeing Like a State: How Certain Schemes to Improve the Human Condition Have Failed*. Yale University Press, New Haven

Sen, S (1999) 'Some Aspects of State–NGO Relationships in India in the Post-independence Era', *Development and Change*, 30: 327–355

Shackleton, S E and B M Campbell (2001) *Devolution in Natural Resource Management: Institutional Arrangements and Power Shifts. A Synthesis of Case Studies from Southern Africa*. SADC Wildlife Sector Natural Resource Management Programme, Lilongwe, Malawi, and WWF (Southern Africa), Harare

Shackleton, S, Campbell, B, Wollenberg, E and Edmunds, D (2002) 'Devolution and Community-based Natural Resource Management: Creating Space for Local People to Participate and Benefit', *Natural Resource Perspective* (ODI), No 76, http://www.odi.org.uk/nrp/76.pdf

Shen M (2001) 'Grain for Green Policy and Its Impacts on Community and Households', Paper presented at the International Symposium Chinese Forest Policy, held in Sichuan Province, June 2001

Shi, T (1999) 'Village Committee Elections in China: Institutionalist Tactics for Democracy', *World Politics*, 51(3): 385–412

Shiva, V 0(1988) *Staying Alive: Women, Ecology and Survival in India*. Kali for Women, New Delhi

Shrestha, N K and C Britt (1997) 'Crafting Community Forestry: Networking and Federation-Building Experiences', Paper presented at the seminar Community Forestry at a Crossroads, Bangkok, Thailand, 17–19 July 1997

Singh, K and V Ballabh (1991) 'People's Participation in Forest Management: Experience of Van Panchayats in UP Hills', *Wasteland News*, August–October, pp5–13

Singh, N M (1995) *Collective Action for Forest Protection and Management by Rural Communities in Orissa*, Paper presented at the Fifth Annual Conference of the International Association for the Study of Common Property, 24–25 May, 1995, Bodoe, Norway

Singh, N M (2000) 'Community Forest Management versus Joint Forest Management: Conflicts between informal self-initiated forest protection efforts and JFM', Paper presented at the International Workshop on JFM, June 2000, New Delhi

Singh, N M (2001) 'Women and Community Forests in Orissa: Rights and Management', *Indian Journal of Gender Studies (IJGS)*, July–December, 8(2): 257–270, Sage Publications, New Delhi

Singh, N M and K Kumar (1994) *Forest Protection by Communities in Sambalpur and Balangir Districts of Orissa*. SIDA, New Delhi, mimeo

Singh, N M and K K Singh (1993) 'Forest Protection by Communities in Orissa – A New Green Revolution', *Forests Trees and People Newsletter*, No 19, Uppsala, Sweden

Singh, S (1997a) *Diverse Property Rights and Diverse Institutions: Forest Management by Village Forest Councils in the U.P. Hills*. IDS, Sussex

Singh, S (1997b) *Collective Dilemmas and Collective Pursuits: Community Management of Van Panchayats (Forest Councils) in the UP Hills*. IDS, Sussex, mimeo

Singh, S (1997c) *Collective Action for Forest Management in the UP Hills*. IDS, Sussex, mimeo

Sioh, M (1998) 'Authorizing the Malaysian Rainforest: Configuring Space, Contesting Claims and Conquering Imaginaries', *Ecumene* 5(2): 144–66

Sivaramakrishnan, K (2000) 'State Sciences and Development Histories: Encoding Local Forestry Knowledge in Bengal', *Development and Change*, 30(1): 61–89

SKS (Sainion Ka Sangathan) (1999a) *Case Study of Gadholi Van Panchayat, District Almora*

SKS (1999b) *Case Study of Anarpa Van Panchayat, District Nainital*

SKS (1999c) Proceedings of NGO discussion held in July, 1999, Bhawali, mimeo

SKS (1999d) *VFJM with Chora Van Panchayat, District Bageshwar*

Somanathan, E (1991) 'Deforestation, Property Rights and Incentives in Central Himalaya', *Economic & Political Weekly*, Bombay, 26 January 1991

Song, Y (1997) 'New Organizational Strategy for Managing the Forests of Southeast China: the Shareholding Integrated Forest Tenure (SHIFT) System', *Forest Ecology and Management*, 91: 183–194

SPWD (Society for the Promotion of Wasteland Development) (2000) *Proceedings of the Seminar on Sustainable Forest Management in Uttarakhand*, 22–23 February, Dehradun

Steins, N and V Edwards (1999) 'Platforms for Collective Action in Multiple Use Common-pool Resources', *Agriculture and Human Values*, 16(3): 241–255.

Suminguit V J, S C Easterluna and M M Paunlagui (2003) 'Creating Space in Kiito, Caayan, Malaybalay, Bukidnon' in A P Contreras (ed) *Creating Space for Local Forest Management in the Philippines*, La Salle Institute of Governance (LSIG) and Antonio Contreras, Manila, Philippines, pp183–198

Sun, C (1992) 'Community Forestry in South China', *Journal of Forestry*, 90(6): 35–40

Sundar, N (1997) *Subalterns and Sovereigns: An Anthropological History of Bastar 1854–1996.* Oxford University Press, Delhi

Sundar, N (1998) 'The Asna Women's Collective: The Interplay of Gender and Environment in a Village Ecological Initiative' in G Persoon and A Kalland (eds) *Environmental Movements in Asia.* Curzon Press, Surrey, pp227–252

Sundar, N (2000a) 'Unpacking the "Joint" in Joint Forest Management', *Development and Change*, 31(1): 255–280

Sundar, N (2000b) 'Is Devolution Democratisation? Evolving State–Society Relations in Forest Management, Bastar, Central India', Draft paper based on research under the CIFOR project Creating Space for Local Forest Management, New Delhi

Tiffen, P and S Zadek (1998) 'Dealing with and in the Global Economy: Fairer Trade in Latin America' in J Blauert and S Zadek (eds) *Mediating Sustainability: Growing Policy from the Grassroots.* Kumerian Press, West Hartford, CT

Tolentino, L L, R F Plopino, and E D V Jacinto (2003) 'Creating Space in Balian, Pangil, Laguna' in A P Contreras, (ed) *Creating Space for Local Forest Management in the Philippines*, La Salle Institute of Governance (LSIG), Manila, Philippines, pp153–164

Torres, C S and F K Mallion (2003) 'Creating Space in Bayagong, Canarem, Aritao, Nueva Vizcaya' in A P Contreras (ed) *Creating Space for Local Forest Management in the Philippines*, La Salle Institute of Governance (LSIG), Manila, Philippines, pp139–152

Touraine, A (1985) 'An Introduction to the Study of Social Movements', *Social Research*, 52(4): 749–787

Upreti, B (2001) 'Beyond Rhetorical Success: Advancing the Potential for the Community Forestry Programme in Nepal to Address Equity Concerns' in E Wollenberg, D Edmunds, L Buck, J Fox and S Brodt (eds), *Social Learning in Community Forests*, Center for International Forestry Research, Bogor, Indonesia, pp189–209

Vasundhara (1997) *Ecological, Institutional and Economic Assessment of Community Forest Management: Village Gadabanikilo, Orissa.* Study done for the Ecological and Institutional Network, National Network for JFM, Bhubaneswar, India, August 1997

Vasundhara (1998) *Effective Local Management of Forests: Learning from self-initiated management organizations in India – Case-Study of Paiksahi Village*, Bhubaneswar, India

Vasundhara (1999) *Report of Broad-based Consultation Process for Pro-poor Forest Policy*, Bhubaneswar, India

Vasundhara (2000) *Study of Federations of Forest Protection Groups in Orissa*, Unpublished draft report, Bhubaneswar, India

Veber, H (1998) 'The Salt of the Montana: Interpreting Indigenous Activism in the Rain Forest', *Cultural Anthropology*, 13(3): 382–413

Vermillion, D L (1997) *Impacts of Irrigation Management Transfer: a review of the evidence.* International Irrigation Management Institute, Colombo, Sri Lanka

Vermillion, D L (1999) 'Property Rights and Collective Action in the Devolution of Irrigation System Management' in R Meinzen-Dick, A Knox, and Di Gregorio (eds) *Collective Action, Property Rights and Devolution of Natural Resource Management: Exchange of Knowledge and Implications for Policy*, Proceedings of the International Conference, Puerto Azul, the Philippines, 21–25 June 1999. DSE/ZEL, Germany, pp183–220

Vira, B (1997) 'Analytical Tools for Assessing Institutional Pluralism in Forestry', Paper prepared for the Workshop on Pluralism, Sustainable Forestry and Rural Development, 9–12 December 1997, FAO, Rome

WCD (World Commission on Dams) (2000) *Dams and Development: A new framework for decision making. The Report of the World Commission on Dams.* Earthscan Publications, London

Western, D, R M Wright and S C Strum (eds) (1994) *Natural Connections: Perspectives on Community-based Conservation.* Island Press, Washington, DC

Wiley, L (1997) *Finding the Right Institutional and Legal Framework for Community-based Natural Forest Management.* CIFOR Special Publication, CIFOR, Bogor, Indonesia

Wollenberg, E, J Anderson and D Edmunds (2001) 'Pluralism and the Less Powerful: Accommodating Multiple Interests in Local Forest Management', *International Journal of Agricultural Resources, Governance and Ecology,* 1 (3/4): 199–222

Wollenberg, E, D Edmunds, L Buck, J Fox and S Brodt (eds) (2001) *Social Learning in Community Forests,* CIFOR, Bogor, Indonesia

Wollenberg, E and A Ingles (1998) *Incomes from the Forest: Methods for the Development and Conservation of Forest Products for Local Communities.* CIFOR, Bogor, Indonesia

Wollenberg, E and H Kartodihardjo (2002) 'Devolution and Indonesia's New Forestry Law' in C J P Colfer and I A P Resosudarmo (eds) *Which Way Forward? People, Forests and Policymaking in Indonesia,* Resources for the Future, Washington, DC, pp81–95

World Bank (1997) *Project Appraisal Document, India, Uttar Pradesh Forestry Project,* Report No 16915-IN, Rural Development Sector Unit, South Asia Region, World Bank, Washington, DC

Yeh, E (2000) 'Forest Claims, Conflicts and Commodification: The Political Ecology of Tibetan Mushroom-Harvesting Villages in Yunnan Province, China', *The China Quarterly,* 161: 264–278

Yin, R (1998) 'Forestry and the Environment in China: the Current Situation and Strategic Choices', *World Development,* 26(12): 2153–2167

Yin, R and D Newman (1997) 'Impacts of Rural Reforms: The Case of the Chinese Forest Sector', *Environment and Development Economics,* 2: 291–305

Zhang, Y (2000) 'Impacts of Economic Reforms on Rural Forestry in China', *Forest Policy and Economics,* 1: 27–40

## In Chinese

Compiling Board of Hubei Provincial Annals of Forestry (1989) *Hubei Provincial Annals of Forestry.* Wuhan Publishing House, Wuhan

Compiling Board of Guizhou Provincial Annals of Forestry (1994) *Guizhou Provincial Annals of Forestry.* Guizhou People's Press, Guiyang, p292

MoF (Ministry of Forestry) (1987) *China Forestry Yearbook 1986.* China Forestry Publishing House, Beijing, pp477–478

MoF (1988) *China Forestry Yearbook 1987.* China Forestry Publishing House, Beijing, pp7–8, 11–12, 485–486

MoF (1990) *China Forestry Yearbook 1989.* China Forestry Publishing House, Beijing, pp37–41, 46–47, 106–132

MoF (1995) *China Forestry Yearbook 1994.* China Forestry Publishing House, Beijing, pp85–105

MoF (1996) *China Forestry Yearbook 1995.* China Forestry Publishing House, Beijing, pp26–29

Wang, L and D Yang (1994) 'A Discussion on Distribution of Economic Benefits from Colletive Forests', *Forestry Economy,* 6: 48–54

Wu, Y (1993) 'Reduce Burden of (ie Tax and Charges on) Farmers to Promote (Forestry) Development', *Forestry Economy*, 1: 48–54

Yunnan Provincial Department of Forestry (1987) *Yunnan Forestry Yearbook 1987.* Yunnan Provincial Department of Forestry, Kunming, pp80–81

# Index